Technology Management in Business

Other Recent Titles of Central West Publishing

Biopolymers Based Advanced Materials
ISBN: 978-0-6482205-4-1 (e-book)
ISBN: 978-0-6482205-5-8 (hardcover)

Polymers in Oil and Gas Industry
ISBN: 978-0-6482205-0-3 (e-book)
ISBN: 978-0-6482205-1-0 (softcover)

Functional Nanomaterials and Nanotechnologies: Applications for Energy & Environment
ISBN: 978-0-6482205-2-7 (e-book)
ISBN: 978-0-6482205-3-4 (softcover)

Functional Polymer Blends and Nanocomposites
ISBN: 978-0-6482205-6-5 (e-book)
ISBN: 978-0-6482205-7-2 (hardcover)

Advances in Polymer Technology: Material Development, Properties and Performance Evaluation
ISBN: 978-1-925823-00-4 (e-book)
ISBN: 978-1-925823-01-1 (hardcover)

Polymer Nanomaterials for Specialty Applications
ISBN: 978-1-925823-03-5 (e-book)
ISBN: 978-1-925823-04-2 (hardcover)

Advanced Materials
ISBN: 978-1-925823-05-9 (e-book)
ISBN: 978-1-925823-06-6 (hardcover)

Technology Management in Business

Dr. Avinash Pawar
Dr. Nitin Zaware
Dr. Kuldip Charak

CWP
Central West Publishing

This edition has been published by Central West Publishing, Australia

For more information about the books published by Central West Publishing, please visit https://centralwestpublishing.com

Disclaimer

NATIONAL LIBRARY OF AUSTRALIA

A catalogue record for this book is available from the National Library of Australia

ISBN (print): 978-1-925823-02-8

About the Authors

Dr. Avinash Pawar is a multifaceted management teacher, trainer and researcher with academic experience across multiple disciplines of management. He has completed PhD (Management), MBA (IT), MBA (HRM), MBA (Operations), MBA (Marketing) and BE (Computer) from Savitribai Phule Pune University. He has a wide range of scholarly interests with an interdisciplinary approach towards management teaching-learning. He is also academically associated with some of the reputed management institutes and has published many research papers in international and national conferences and journals. He was awarded with Best Teacher Award, Outstanding Management Teacher Award and Global Management Teacher Award by prestigious educational organizations.

Dr. Nitin Zaware is a distinguished professor of management and has vast teaching and research experience in management. He has contributed more than 70 research papers on various management topics in research journals, national and international conferences. He has also been a member of prestigious professional bodies like Indian Society for Technical Education, National Institute of Personnel Management, Strategic Management Forum, All India Commerce Association and Association of Indian College Principals. He is also a member of the Board of Studies and a PhD research guide at Savitribai Phule Pune University.

Dr. Kuldip Charak is an eminent professor of OB and HRM. He has immense experience in finance, administration, HR, training and development, besides highest qualification in the field of education. He worked in top management and has always strived to produce qualitative product and services in public and private organizations. He has obtained PhD from University of Pune and has unique expertise in developing advanced learning methodologies for management students. He is the member of the Board of Studies and a PhD research guide at Savitribai Phule Pune University. He has published many research papers and articles in leading journals and conferences.

Dedication

It gives us great esteem and pleasure to publish this book titled 'Technology Management in Business' with Central West Publishing Australia.

We are obliged and thankful to all our colleagues and friends in academic fraternity.

Last but not the least, we have no words to express our deepest feelings for the support, encouragement and blessings from our beloved parents and family members.

Thank you

Dr. Avinash Pawar
Dr. Nitin Zaware
Dr. Kuldip Charak

Contents

Preface

In an endeavor to continue improvement in domain knowledge in the areas of business, technology and management, we have written this book to provide a framework for learning in a way that emphasizes the uniqueness of integrating and implementing technology in businesses.

In a globalized competitive environment, technology has a crucial role in determining success and growth of the organization and plays very decisive role in gaining and sustaining competitive advantage for any business. Technology has become inescapable and unavoidable in managing any organization as it has direct impact on costs and profitability of the business. Hence, effective management of technology is essential for both product and service organizations. The importance of management of intellectual property rights in managing technology is also growing.

This book has the purpose of developing understanding of technology management with emphasis on strategic decisions involved in it, along with describing the important techniques used in the management of technology. Thus, the book focuses on the basic concepts and processes involved in all major components of management of technology. In addition, it covers both macro and micro issues involved in technology management.

We are thankful to our teachers, friends in academic fraternity and our students for providing us this opportunity to contribute to the important area of Technology Management in Business. We hope that readers will find it easy and interesting to learn and understand various concepts and processes described in the book.

Dr. Avinash Pawar
Dr. Nitin Zaware
Dr. Kuldip Charak

1

Overview of Technology Management

1.1 Introduction

Technology has become an integral and essential part of not only businesses but of individual life. Technology has a profound effect on all aspects of life: business, economy, politics, society and economics. One school of thought considers technology to be a determining factor, which is almost shaping direction/trends in all aspects of society, business, behavior, etc. In 1998, Microsoft's Bill Gates said about the Wright brothers' invention at Time Magazine's 75th anniversary celebration of the airplane: "We have to understand that engineering breakthroughs are not just mechanical or scientific, they are liberating forces that can continually improve people's lives."

In today's business environment, poor technology management can be certainly a reason of failure of businesses. On the contrary, technology itself is a competitive advantage for many organizations. The speed of development of new technology and processes necessitates a structured approach to managing technology. Technology facilitates fulfilment of customers' need for products and services efficiently and effectively. In many businesses, providing customer service without the use of high technology has become impossible. Similarly, use of technology has become necessary for all business processes to remain competitive in business. In 'Technology Management', strategic decisions related to technology are made taking integrated organization wide approach to meet business objectives effectively. The overall scope of technology management is, thus, much beyond information technology management. It considers using technology in all processes to design a product or service, to manufacture a product, to make product/service available so as to increase profits, satisfy customers, increase business and stay competitive.

In this unit, a basic understanding of technology, its evolution, industrial revolution and management of technology has been provided. The major aspects of technology management are discussed in the subsequent units.

1.2 Concept of Technology

Technology is human innovation that involves the application or generation of knowledge or processes to develop systems that solve problems or extend human capability. Technology itself is very old. Answer to the question "how old is technology?" will vary from 2.5 million years to 300 years depending on the responders' perception of technology. 'Technology' in English language is a word with origins in the Greek word technologia, which is a combination of techne meaning 'craft' and logia meaning 'saying'. Technology literally means the act or study of making or crafting. It refers to the collection of processes and knowledge that people use to extend human abilities and to satisfy human needs and wants. Some of the definitions of technology are shown below:

- A body of knowledge used to create tools, develop skills and extract or collect materials; the application of science (the combination of the scientific methods and materials) to meet an objective or solve a problem.
- The practical application of science to commerce or industry engineering: the discipline dealing with the art or science of applying scientific knowledge to practical problems.
- The application of scientific advances to benefit humanity.
- Anything people have made could be called technology, but technology usually implies advancement in using scientific knowledge and applying it to a practical use.

It can be seen that in all the above definitions, purpose of technology is in variance. We will consider a holistic and broad based definition of technology: Technology is a broad concept that deals with a species' usage and knowledge of tools and crafts, and how it affects a species' ability to control and adapt to its environment.

1.2.1 Aspects of Technology

Technology influences the society and culture where it is developed and used. Technology has three major aspects which are interrelated and mutually reinforcing. These aspects are: products and services, human activities that create these products and services as well as skills/capabilities that enable technical activities. These three aspects are shown in Figure 1.1.

Society's demand for products and services stimulates development of technology and products/services. This expands technical capabilities and skills. Research and development activities create new products/services to the society. These activities also enable developing skills/capability in technology.

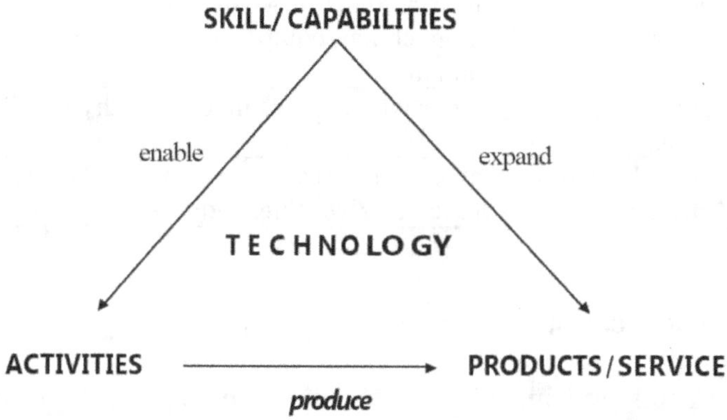

Figure 1.1 Aspects of technology.

1.2.2 Impact of Technology

Technology has affected all aspects of our life and environment. Our life has become fully dependent of technology so much so to call us the slaves of technology. The impact of technology can be classified to be in these areas:

- Society including behavior
- Culture
- Economy
- Environment
- Politics
- Ethical considerations

1.2.3 Classes of Technology

To gain insight of technology, it is desirable that technology should be understood in terms of different classes. Technology can be broadly categorized in the following four classes, as shown in Table 1.1.

Table 1.1 Classes of technology

No.	Class of Technology	Particulars
1	Technology as Knowledge	The knowledge, learning, know-how behind technological innovation, development
2	Technology as Objects	Physical item/device of technical performance, e.g. machines, equipment, instruments, weapons, appliances
3	Technology as a Process	Begins with a need and ends with a solution
4	Technology as Activities	Set of technical activities, their methods, procedures, routines, skills, etc.

1.2.4 Resources of Technology

1. Tools and machines – Machines are of any type useful to make parts, process parts, and make tools.
2. Materials - Raw materials that are transformed into useful products.
3. Processes - Processes are the tasks/jobs through which products are produced.
4. Energy - Energy is the source of power.
5. Information - Technology requires information. It is the use of the information that helps us to meet our needs and wants.
6. Humans - People are the creators of technology and users of technology. People use many other resources to create technology.
7. Capital - Any form of wealth.

1.3 Growth of Technology

Technology is a material entity created by the application of mental and physical effort to nature in order to achieve certain value. From this definition, technology is very old, as humans have been using primitive tools for almost 2 million years. However, making objects that go slightly beyond this took a while to develop. In this aspect, the technology is approximately 500,000 years old. In general, the technology evolves in three stages of tools, machine and automation, as shown in Figure 1.2.

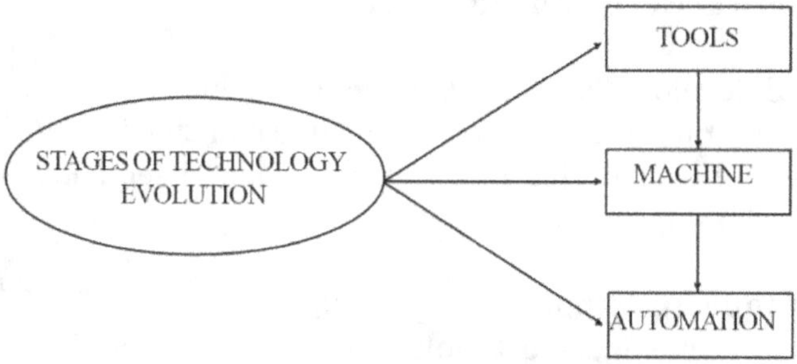

Figure 1.2 Stages of technology evolution.

A) Tool: Tool is the first stage of evolution of technology. A tool makes a task simple or more efficient or reduces physical effort by providing a mechanical advantage in accomplishing a physical task. Tool is used by applying human or animal effort. The first tools developed by humans must have been for food and safety. For instance, hunters developed tools for hunting. Spear, arrow, container, plow or hammers are some of the examples of tools. Initially, all the tools developed were deployed by the hunters/gatherers. Subsequently, animal-powered tools such as the plow and the horse were developed. Use of animal driven tools increased productivity and output in food production.

B) Machine: The second technological stage was the development of the machine. Machine substitutes human physical effort. A machine requires the operator. The operator's function is only to control. With the industrial revolution, machines became widespread. Machines work far beyond the capacity of human beings. Machines also perform seemingly impossible tasks due to limitations of physical and mental capacity. For example, a simple drilling machine can drill 20 times more than manually achievable. Besides, manual drill can never achieve the quality a hole drilled on a drilling machine provides.

C) Automation: Automation is the third stage of technological evolution. The automation is a device that removes the element of human control with an automatic algorithm. For example, an automatic watch does not require winding up the spring or adjusting time. All the three stages of evolution continue to be relevant today. We continue to use tool, machine and automation devices.

1.4 Historical Developments in Technology

Historical development of technology is the history of development of man from the cave in the forest to an urban city. For convenience, development is classified into seven periods in sequence. Development at each stage is stated in brief.

1. **Stone Age** (approximately 250,000 B.C. - 3,000 B.C.): During the Stone Age, most tools were made of stone. Shaping rocks into sharp edges for arrows, and knives was time consuming and the tools would easily wear with use. Most tools were used for agriculture and hunting.
2. **Bronze Age** (approximately 3000 B.C. - 1200 B.C.): Bronze is a soft metal alloy. It is made of copper and tin. It is easier to shape tools if metal was melted and then shaped by using a mould. The metal tools were preferred by people because these were more durable and useful than stone tools. Besides, these were easier to shape.
3. **Iron Age** (approximately 1200 B.C. - 500 A.D.): Iron metal is stronger than bronze when alloyed with other materials. Again, most technological developments used during the period were brought about to improve agriculture, trade and military weapons.
4. **Middle Age** (approximately 500 A.D. - 1500 A.D.): The Middle Age brought about a number of technological developments. This has resulted in industrialization. The agricultural development led to production of a surplus of crops and increased trade. Increased trade created bigger markets with more products.
5. **Renaissance** (approximately 1450 A.D. - 1700 A.D.): The Renaissance was a time of rebirth in the arts. Gutenberg invented the printing press. This facilitated information to be disseminated throughout the world. Initially, the printing press was used for printing the holy Bible. Subsequently, it was used to print other literature. The first screwdriver was invented during this time period. The camera, telescope, submarine and hydraulic press were some other products developed during this time period.
6. **Industrial Age** (approximately 1700 – 1940): Industrial Age was the beginning of factory system. Factories were increasing in numbers rapidly. The growing number of factories

created the need for technological improvements in products and processes. International trade increased which created a bigger demand for efficient, fast and reliable systems for transportation of products. Communication advances accelerated the information and coordination systems at an alarming rate. Structural systems were under tremendous stress to improve as bigger cities began to grow up instead of out.

7. **Information Age** (approximately 1940 – present): The Information Age is the current period of time. During information age, technological developments take place at an exponential rate. Product life is shortened. The microchip revolutionized the world of electronics. It has made communication systems faster, cheaper and more powerful. Research in energy systems continue to make them more efficient, user friendly as well as less harmful to the environment.

1.5 Management of Technology

Gerald H. Gaynor, the famous author of Handbook on Technology Management, has provided a comprehensive definition of technology management as: "Managing technology is a method of operation that leverages human resources, technology and other business assets by optimizing the relationships between the technology functions of the business enterprise. It is the process of integrating science, engineering and managing with research, development and manufacturing in order to meet the operational goals of the business unit effectively, efficiently and economically. It includes managing the totality of the technology operations from concept through commercialization." Thus, the scope of technology management is quite vast and requires an integrated approach involving various functions. Technology management can be defined in the same manner as any other management functions. Following are the functions of management:

- Planning
- Organizing
- Staffing
- Implementing
- Controlling/Evaluation

Considering the above functions in context of technology, it becomes Technology Management (TM). In other words, technology management can be defined as a process of planning, organizing, staffing, implementing and monitoring/evaluation technology to achieve the stated objectives.

Technology management embraces several interconnected issues such as: technology policy; technological forecasting and assessment; technology strategy; technology transfer; technology-induced as well as market-oriented research and development (R&D); process technology and product technology and their continuing improvement; human resource management in terms of innovative capabilities, flexibility and contribution; and technology project management. TM was set into motion when man invented the wheel, except that it was never practiced consciously. Now, TM has become what it never was before, an organized and systematic discipline. As technology management embraces several interconnected issues ranging from policy planning at the national level to strategic planning at the firm level, it calls for decisions and result-oriented actions at the macro-as well as micro-levels and an effective macro-micro linkage.

Macro technology management commonly refers to technology management at the national level. It includes:

- Planning for the development of technological capabilities at the national level.
- Identification of key sectoral technology and related fields to be developed.
- Determining 'make' or 'buy' decisions, i.e., whether importation or self-development is to be pursued.
- Establishment of institutional mechanisms for directing and coordinating the development of national technological capabilities.
- Design of policy measures for controls.

Micro technology management concerns technology management at the firm or project level. It includes:

- Responding to competitors who are using technology as a strategic weapon.
- Integrating technology strategy into the overall corporate strategy.

- Identifying and evaluating various technological options and innovations and the factors relating to their success and failure.
- Directing research and development itself, including determination and definition of project feasibility.
- Monitoring and planning technological obsolescence and replacement.

Both macro and micro-technology management seek to raise economic efficiency. Micro TM is the basis for macro TM, while the latter provides guidelines and an environment for the former. Consistency among these two levels of management is essential, but institutional mechanisms will largely determine whether they are effectively combined. Though macro-support could catalyze changes, the real actions have to take place at the industry level.

1.6 Functions of Technology Management

Technology management embraces many functions which are interrelated to each other and are specialized. Some of the functions of technology management are mentioned below. These functions will be discussed in subsequent units.

- Technology policy
- Technology forecasting
- Technology strategy
- Technology transfer
- Technology absorption
- Technology diffusion
- Technology acquisition
- Research and development management
- IPR management
- Technology assessment
- Innovation management
- Environment impact analysis

1.7 Significance of Technology Management

Management of technology is of critical significance to the organizations operating in globalized and competitive business environment. Technology is unavoidable and inescapable. In a competitive business environment only those organizations survive who focus their efforts towards customer satisfaction. Accordingly, all business processes are reengineered towards customer satisfaction. Meeting customer requirement is the basic to satisfy customers. Customer requirement in generic terms can be stated as follows:

- Quality: Customer wants better and more in quality. Customer expects more features and better performance of existing features. One of the examples is cell phones. Over the years, quality in terms of features and performance is showing rapid improvements.
- Delivery: Customers expect faster delivery, i.e., delivery in lesser time period.
- Variety: Customer expects more variety in the products/services. The manufacturer should be able to design/develop more products and introduce them in the market quickly.
- Service: Customer will ideally like to have product without need of any service during its lifetime.
- Cost: Customer would like to have all of the above and more, but is unwilling to pay anything for it. On the contrary, the customers expect the prices to reduce.

Technology is the most potent weapon by which an organization can achieve the above. By a systematic and planned approach for managing technology, an organization can bring about drastic improvement in performance, which may not be possible otherwise. By proper decisions on managing technology, an organization can achieve the following:

- Faster and economic design
- Better quality of products
- Increased sales
- Better customer satisfaction
- Customer loyalty
- Higher profits
- Higher value
- Lesser costs
- Better resource utilization
- Lesser throughput time
- Higher customer service and more

Besides, by proper management of IPR (intellectual property rights), an organization can increase its revenue substantially and ward off competition. Effective management of technology development would facilitate higher success of high-tech products on launch.

A) Product Technology: Product technology is concerned with application of technology in the product/services. This involves creating a product/service by using technology to fulfil customer needs. Sony, after extensive research developed an innovative product, Headphone Stereo Walkman TPS-L2 and launched in 1979. Sony's LCD and 3D Television are also examples of application of technology for creating products. Similar example is introduction of NANO car by Tata Motors. As we know, this was the world's first low cost car with an innovative design. Another example of product technology is popular car Model T developed several decades earlier. More than 3 million cars of model T were sold. Penicillin is the first antibiotic drug for treatment of diseases caused by microbes. Without penicillin, 75% of the people now alive would not be alive because their parents or grandparents would have died due to bacterial infections. The effects of a drug like this are absolutely mind-boggling. This is one of the excellent examples of application of technology leading to new breakthrough products. Before penicillin, if one had a serious infection, he/she would simply die. Newer antibiotics for all the three, gram positive, gram negative and both (broad spectrum) are being developed. These are necessary as a response to immunity of earlier antibiotic drugs due to excessive consumption and reduction in side effects. Patients with advanced diabetes can't use the energy stored in their bodies as their bodies stop making a hormone known as insulin. It is needed to convert sugar into energy. Canadian researchers Frederick Grant Banting and Charles Best first identified insulin in 1921. In 1922, a Canadian patient received the first successful treatment with insulin extracted from an animal. Insulin has completely changed the lives of diabetes patients. Swann says, "People with diabetes didn't have much chance for a long life. Insulin has made the life better for the diabetics." Smallpox is one of the most dreaded diseases. Due to development of smallpox vaccination, it is the first disease to be wiped from the face of the earth. So is the case of polio, which is on the verge of being the second scourge to be eliminated due to the development of highly successful vaccine for the disease.

B) Process Technology: Process technology is concerned with how to make a product. One can refer to the examples of technological development during industrial revolution. For example, John Roebuck developed a new technology for manufacturing of sulfuric acid called Lead Chamber process. By using the process, sulfuric acid in large quantities could be produced at lesser cost. Similarly, Nicolas LeBlanc developed LeBlanc process for the production of alkali in large quantities. In these cases, the products in terms of attributes and features have not changed, but new technology of making these products was deployed. The application of technology in processes have following benefits:

- Reduced processing time and better safety
- Better quality and reduced rejections
- Lesser cost, Less fatigue and efforts

An organization should constantly improve its processes by application of technology. At the same time, based on the need, it should constantly improve its products. Organizations need to be proactive rather than reactive to face the challenges of competition.

1.8 Technological Change

Technological change has been broadly defined as 'the process by which economies change over time in respect of the products and services they produce and the processes used to produce them'. More specifically, it has been termed as 'alteration in physical processes, materials, machinery or equipment, which has impact on the way work is performed or on the efficiency or effectiveness of the enterprise'. Technological change may involve a change in the output, raw materials, work organization or management techniques, but in all cases, it affects the relationship between labor, capital and other factors of production. As the policies to stimulate technological progress and productivity growth - at both national and firm levels - must be formulated in a broad socio-economic context, their focus ought to be on the internal dynamics of technological change. It has been suggested that the knowledge pertaining to technological change in the less developed countries would be a crucial input to the understanding of the phenomenon in industrialized nations. In this light, the following paragraphs provide a selective survey of theories and measurement of technological change.

There are following theories of technological change:

A) Neo-Classical Theory
B) Marxist Theory
C) Schumpeter's Theory
D) Evolutionary Theory
E) Market-Pull Theory
F) Technology-Push Theory

A) Neo-Classical Theory: The basic tool for the study of technological change is the notion of a production function which specifies a quantitative relation between inputs and outputs. The most common inputs are capital and labor, which are called factors of production. The production function can be represented as a series of isoquants - curves corresponding to the constant output obtainable by the infinite number of available combinations of the factors of production. At any given time, there is a given level of technology which determines the techniques available for production. According to this theory, technological change takes place in the form of shifts of the production function towards the origin.

B) Marxist Theory: Karl Marx perceived technology as not self-generating, but as a process directed by willful, conscious, active people and molded by historical forces. He held that technological change - the development of the productive forces - was the prime mover of history. The individual entrepreneur invests and innovates because it is rational for profit maximization or necessary for survival. Marx seemed to be under the spell that innovations simply must be labor-saving.

C) Schumpeter's Theory: This theory views innovation as the engine of economic development and as a disequilibrium phenomenon. Innovation is defined as the carrying out of new combinations of means of production, which include a wide variety of cases such as: the introduction of a new good or of a new quality of a good, or of a new method of production, the opening of a new market, the conquest of a new source of supply of raw materials, the carrying out of a new organization of any industry, etc. The emphasis is laid on the notion that technological change is to be understood as a case of innovation more generally and not as another piece of routine economic behavior. Schumpeter's formulation of production function differed from neo-classical theory in that capital was excluded and only labor and land were included as inputs.

D) Evolutionary Theory: This suggests a biological analogy to explain technological change. The Darwinian two-state process of mutation (invention) and selection (innovation) has been employed to understand the evolution of technology. Biological evolution appears to have a certain correspondence with the interpretation of technological changes in industrial sectors - from a state of flux when product innovation prevails in the search for a successful design, to a maturity phase where incremental process innovation prevails.

E) Market-Pull Theory: Markets govern the innovation process. The market constitutes a communication channel through which political, economic, social and ecological forces influence buyers in their demand for technological products. Continuous changes in these forces have an impact on the response provided by technology with respect to the type, capabilities, performance, safeguards, solutions, etc. These messages are transmitted and communicated through the market where buyer's requirements (themselves influenced by external forces) are matched with technological changes and where future demands can be detected by the producers of technology.

F) Technology-Push Theory: Technology is defined as an autonomous or quasi-autonomous factor. It assumes a one-way causal determination approach, i.e., from science to technology to the economy. It proposes that technological developments occur exogenously through discoveries, theories, ideas as well as research and development work, which may or may not then create (or be matched with latent) demand for their output.

2

Competitive Advantages through Technologies

2.1 Introduction

Competition is a contest between individuals, groups, animals, etc., for territory, a niche or a location of resources. It arises whenever two or more entities strive for a goal and objective which cannot be shared. In business, it is often associated with competition as most companies are in competition with at least one other firm over the same group of customers. In other words, it is the existence within a market for some good or service of a sufficient number of buyers and sellers such that no single market participant has enough influence to determine the going price of the good or service. The phenomenon is exactly opposite of monopoly. Competition is nothing but a rivalry in which every seller tries to get what other sellers are seeking at the same time: sales, profit and market share by offering the best practicable combination of price, quality and service. As the market information flows freely, competition plays a regulatory function in balancing demand and supply.

2.2 Competitive Advantage

It is defined as the strategic advantage one business entity has over its rival entities within its competitive industry. Achieving competitive advantage strengthens and positions a business better within the business environment. As a firm sustains profits that exceed the average for its industry, the firm is said to possess a competitive advantage over its rivals. The goal of much of business is to achieve sustainable competitive advantage. The concept of competitive advantage lies at the heart of understanding a firm's performance in competitive markets; above average performance in the long-run can only be generated by creating a sustainable competitive advantage.

Cost and differentiation advantages (Figure 2.1) are referred as 'positional advantages' as they describe firm's position in the industry/sector as a leader in either cost or differentiation. Resource based view, i.e., a firm utilizes its resources and capabilities to create a competitive advantage, ultimately results into superior value creation. To understand the 'competitive advantage', it is necessary to refer to Michael Porter's views and his 'Theory of Competitive Advantage'.

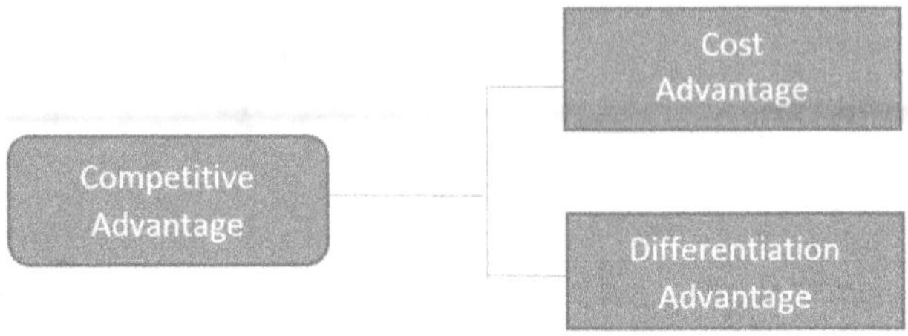

Figure 2.1 Cost and differentiation advantage.

Competitive advantage is based on the theory that seeks to address some of the criticisms of comparative advantage. To attain competitive advantage, an organization has to opt for analysis of environment. It consists of following:

- **External Analysis:** Analyzing the external environment to identify the opportunities and threats. Environment comprises of generic environment and specific/industry/competitive environment.
- **Internal Analysis:** Analyzing the internal aspects of organization to identify the strengths and weaknesses.

2.2.1 PESTEL Analysis

A) Political (Government)

- Political uncertainty (impact of elections)
- Capitalism v/s communism, state v/s central
- State policies (taxes, octroi, power, incentives, subsidies, concessions, etc.)
- Government as regulatory mechanism
- Anti-trust laws and taxation laws
- Deregulation philosophies

B) Economic

- Government's economic policies – BOP, balanced regional development, monetary, fiscal, EXIM policies
- Role of FI, financial markets
- Savings rate – personal/business

- Standard of living/quality of work life
- Interest rates, inflation rate
- Growing international interdependence
- Unemployment
- Trends in GDP
- Changes in stock market valuations
- Trade deficit/surplus

C) Social

- Protests by local residents (Bhumi-Putra)
- Social values v/s organizational values
- Aging population
- Changes in ethnic composition/diversity
- Geographic distribution of population
- Greater disparities in income levels
- More women in the workforce
- Increase in temporary workers
- Postponement of family formation
- Shift in work/career preferences
- Shift in product/service preferences
- Income contribution

D) Technological

- Technology absorption, adoption, dispersal
- Evolution of IT/internet/wireless technology
- Miniaturization of computing technologies
- R&D orientation, industry-academic relations
- Automation, computer-aided design/computer-aided manufacturing systems
- Research in synthetic and exotic materials
- Major technological change (HMT v/s Titan), (cement wet – dry), (conventional TVs v/s LCD, LED, plasma and 3D), (telex - fax – scanners -internet), (landline phones v/s cell phones), (computers: desktops – laptops– iPad – note pad)
- Product innovations
- Government support for R&D

E) Ecological/Environmental

- Pollution – greenhouse effect, etc.
- Global warming, El Nino, etc.
- Enhanced environment degradation

- Legislation to protect ecological balance

F) Legal (Legislations/Statutory)

- Competition act, FEMA, dis-investment act, SICA, pollution control act, essential commodities act, etc.
- Deregulation of utility and other industries
- Taxation at local, state, central levels
- Legislation on corporate governance reforms

2.2.2 Porter's Five Forces Model

Michael Porter proposed the theory (Figure 2.2) in 1985. Competitive advantage theory suggests that states and businesses should pursue policies that create high-quality goods to sell at high prices in the market. Porter emphasizes productivity growth as the focus of national strategies.

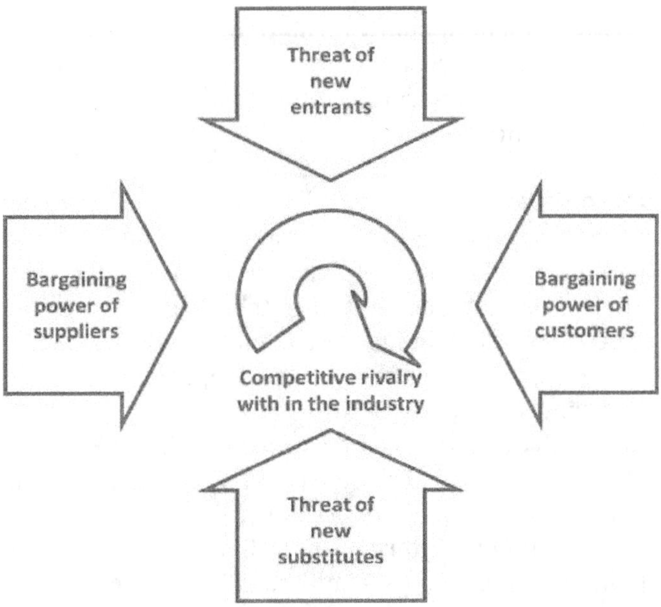

Figure 2.2 Porter's Five Forces model.

Competitive advantage rests on the notion that cheap labor is ubiquitous and natural resources are not necessary for a healthy economy. Porter's ideas about competitive advantage can be used to examine how the information systems affect the performance of a business organization by changing the relationships within the five forces that shape its competitive environment. Porter defines five factors, which determine industry profitability and the extent of the competitive forces faced by organizations. These factors should be considered by any firm that is assessing its position as part of the development of a strategic plan.

1) Threat of New Entrants

It determines the likelihood that new firms will enter the industry and compete away the value either passing it on to the buyers in the form of lower prices or raise the cost of competing. Threat of entry will place a limit on prices and shape the investment to deter entrants. Porter lists many barriers to enter:

- Economies of scale
- Product differentiation
- Capital requirements
- Switching costs
- Access to distribution channels
- Government policy

The height of the entry barriers will make it difficult for a new entrant by forcing him to enter by bidding down a low price.

2) Bargaining Power of the Buyers

Power of buyers can also influence cost and investment, since powerful buyers demand costly service. Porter found that power of a buyer group seems to be strongest, when the following conditions apply as below:

- It is concentrated or has purchased large volumes relative to seller sales
- The products it purchased represent a significant fraction of the buyer's cost of purchases
- The products it purchases from the industry are standard or undifferentiated
- It has faced few switching costs
- It earns low profits
- Buyers pose a credible threat of backward integration
- The industry's product is unimportant to the quality of the buyer's product

- The buyer has full information

Satisfying buyer needs is at the core of success in business. Buyers must be willing to pay a price for a product that exceeds its cost of production; otherwise the industries cannot survive in the long run.

3) Threat of Substitute Products/Services

Threat of substitute products influences the prices a company can charge. Porter suggests that those substitutes, which should not be monitored closely, are:

- Those products, which are providing a better performance and price standard than the industry standard
- Products produced by industries earning high profits
- The threat of substitutes determines the extent to which some other product can meet the same buyer needs and, thus, place a ceiling on the amount a buyer is willing to pay for an industry's product

4) Bargaining Power of the Suppliers

Bargaining power of suppliers determines the cost of raw material and other inputs. According to Porter, the main determinants of suppliers having power over an industry occur when:

- It is dominated by a few companies and is more concentrated than the industry to which it sells
- It is not obliged to contend with other substitute products for sale to the industry
- The industry is not an important customer of the supplier group
- The supplier's products are differentiated or it has built up switching costs
- The supplier poses a credible threat of forward integration

5) Rivalry among Existing Firms

Intensity of rivalry will influence prices as well as certain other areas like advertising, sales promotion, production development, etc. According to Porter, rivalry is intensified by the following factors:

- Numerous or equally balanced competitors
- Slow industry growth

- High fixed or storage costs
- Lack of differentiation or switching costs
- Capacity augmented in large increments
- Diverse competitors
- High strategic stakes
- High exit barriers

The intensity of rivalry acts similarly to the threat of entry. It determines the extent to which firms already in an industry will compete away the value they create for buyers by passing it on to the buyers in lower prices or dissipate it in the higher cost of competing. It plays a major role in determining whether existing firms will expand capacity aggressively or choose to maintain profitability.

The Essence of Five Forces

The strength of these five forces varies from firm to firm. In some industries, all the five forces may be favorable. In some, the pressure from one/more of the forces could be intense. Thus, these five competitive forces will influence price, cost, investment, ROI, etc. Firms through their strategies can influence these five forces. The collective strength of the above mentioned five forces determine the ultimate profit potential of an industry. The strategist's goal should be to find a position in the industry where his company can best defend itself against these five forces or can influence them in his company's favor. The strategy should be able to influence all these five forces in favor of the business, so that the business can secure for itself a position of superiority in the market in relation to competition.

2.3 Technology and Competitive Advantage

Technological change has an important influence on competitive advantage both because it creates new opportunities for competition and because it plays a central part in the existing competitive strategy through its ubiquitous presence in the value chain. Information technology and information systems are very important as every activity creates and uses information. Modern information system technology plays a particularly crucial role in scheduling, controlling, optimizing, measuring and coordinating all manner of activities.

Technological innovations can have important strategic implications for individual companies and can greatly influence industries as a whole. Yet, not all technological change is strategically beneficial. Few years earlier, customers/consumers were buying products and/or availing services either for 'need' or for 'want' by way of technological innovation (need – as required, want – either to create or to maintain status/prestige/ego/image in the society). On account of globalization, since last few years, markets are customer driven. To remain in fierce competition by way of attracting the customers/consumers by satisfying their 'needs' and 'wants' simultaneously, organizations are already working with reverse engineering process, i.e., apart from 7 Ps of marketing process, one more element – utility of the product or service - is added to the entire mechanism. The activity is based on competitive price platform. For example, technological innovation has delivered:

- Multi-sim card mobile phones
- Introduction of small cars, i.e., Tata Nano
- LCD/LED/plasma/3D televisions
- Take-away convenience associated with food items/consumable products
- Introduction of handy laptops

Taking an example of developing country like India, we can refer to the launch of 'Nano', the smallest car with cheapest price tag by Tata Motors. Prior to it, none of the auto major had given a thought to design-develop-manufacture and deliver such concept car in the market. However, with reference to 'Blue Ocean Strategy', Tata Motors started working on the concept car platform and finally delivered the same. Now to beat the competition, almost every auto major has introduced small car models (though they are not in-line with that of Nano). For instance, Maruti-Suzuki and Hyundai have already developed a proto-type of the car which will compete with Nano. Recently, Honda has introduced a small car 'Brio' and Hyundai has also launched its small car 'Eon'. Above quotes are the examples of new technology concepts and innovative approach which will definitely meet customer's requirements in terms of 'wants' and 'needs'. Even in consumer electronics, we can refer to introduction of multi-sim card mobile phones with features, sizes and many more by different manufacturers at competitive prices. This move has already benefited competitive advantage for some of the brands especially new ones, viz. Micromax, Spice, Carbonn, Fly, etc., over their more established market rivals. Since last one year or so, the erstwhile market leaders in mobile phone industry have been observed to lose market share to a great extent due to various reasons in which 'technology' is one of the factors. It has been noticed that some of their models introduced in the last 2-3 years were unable to sustain with the competition due to lack of technological aspect.

2.4 Government and Competitive Advantage

It is clear that to achieve competitive advantage, every businessman has to focus on technology upgradation associated with his product and/or service. As such, it is necessary to ensure contribution by each and every element that is the part of process called as 'competitive advantage'. Let us try to understand the concept by referring to the progress of developing country like India. India is one of the prime members who are actively participating in the globalization process since 1991. Today, every entity has no choice but to compete with at least one member in the said business/process stream. The government using various channels is supporting industries at all levels through the ways mentioned below:

- Financial support – to ensure participation of financial institutions at national level as well as to call for international financial institutions.
- Intellectual support – to provide guidance/consultancy in the related sector.
- Incentives – an opportunity for existing businesses to enter into global business market for the purpose of establishing 'brand image' as well as to earn valuable foreign exchange for the country by way of exports.

In addition, the technical support is provided to:

a) support the industries to opt for either joint venture (JV) or technical collaboration for upgradation of existing product line as well as to introduce new concepts based on latest technological innovations ultimately to satisfy the final customer in terms of value.
b) call for latest and/or most suitable technology available across the globe.
c) invite 'foreign direct investment (FDI)' by offering incentives/rebates on expatriation basis.
d) introduce various export/import substitution schemes.

Department of Science & Technology (DST) of Government of India is playing a crucial role in the technological support process. The Ministry of Science and Technology is the central authority in India responsible for overall development of science and technology as well as for promotion of innovation process in India. DST is the nodal department of the ministry for organizing, coordinating and promoting science and technology (S&T) activities in the country. It is the main public policy support body for science and technology. It has enrolled into a role of strengthening the research and development (R&D) base of the country. It is also the major agency in the country for promotion of basic sciences with a share of nearly 50% of extra mural funding of research projects. The department plays a critical role in the technology development programs relating to key sectors and in the development of international cooperation in science and technology, popularization of science, relating science to society, promotion of entrepreneurship, etc.

National resources data management and generation of databases and initiatives targeting empowerment of women through applications of science and technology form important areas of focus for planned interventions by the department. The department is actively involved in promotion and development of S&T in the states as well as in supporting S&T entrepreneurship development for promotion of knowledge-based technology driven entrepreneurs. It has been supporting an array of autonomous institutions engaged in basic research, technology forecasting, laboratory accreditation and science popularization. Output indicators of these institutions reveal a strong correlation between inputs and outputs. The department oversees two sub-departments, namely, Survey of India and National Atlas and Thematic Mapping Organization (NATMO).

DST has also set up National Innovation Foundation (NIF) of India, with the main goal of providing institutional support in scouting, spawning, sustaining and scaling up grassroots green innovations and helping their transition to self-supporting activities. NIF seeks to achieve this goal by drawing upon the honeybee network and its collaborating partners. The honeybee network and Society for Research and Initiatives for Sustainable Technologies and Institutions (SRISTI) have been scouting innovations by farmers, artisans, women, etc., at the grassroots level. The honeybee database of 10,000 innovations, collected and documented by SRISTI, would be a part of the National Register of Innovations, and is to be managed and supported by NIF.

The primary objectives of NIF are:

- To help the country become an innovative and creative society and a global leader in sustainable technologies by scouting, spawning and sustaining grassroots innovations.
- To ensure evolution and diffusion of green grassroots innovation in a selective, time bound and mission oriented basis so as to meet the socio-economic and environmental needs of the society.
- To provide institutional support in scouting, spawning, sustaining and scaling up grassroots green innovations as well as outstanding traditional knowledge and helping their transition to self-supporting activities.
- To build linkages between excellence in formal scientific systems and informal knowledge systems and create a knowledge network to link various stakeholders through application of information technology and other means, etc.
- To meet its objectives in scouting, awarding and incubation of grass-root innovations.

NIF has set up five dedicated departments to execute innovation to enterprise development models:

- Scouting and documentation
- Business development and micro venture
- Value addition; research and development
- Intellectual property management
- Dissemination and information technology management

Another main department of Ministry of Science and Technology is the Department of Scientific and Industrial Research (DSIR) which has a mandate to carry out the activities relating to indigenous technology promotion, development, utilization and transfer. The primary endeavor of DSIR is to promote R&D by the industries, support a larger cross section of small and medium industrial units to develop state-of-the art globally competitive technologies of high commercial potential, catalyze faster commercialization of lab-scale R&D, enhance the share of technology intensive exports in overall exports, strengthen industrial consultancy and technology management capabilities as well as establish user friendly information network to facilitate scientific and industrial research in the country. It also provides a link between scientific laboratories and industrial establishments for transfer of technologies through National Research Development

Corporation (NRDC) and facilitates investment in R&D through Central Electronics Limited (CEL). It has two autonomous organizations like Council for Scientific and Industrial Research (CSIR) and Consultancy Development Centre (CDC). Hence, the main objectives with which DSIR has been set up are to:

- Promote R&D in industry.
- Develop technological capabilities of technical personnel in national labs/institutions and industry.
- Promote technology development and management in industry.
- Help individual innovators promote original ideas and translate into workable processes/prototypes.
- Promote association of national labs and national institutes with industrial projects for technology development.

Since 1991, i.e., post globalization era, a tremendous activity has been noticed that has taken place in almost every sector, viz.

- Automobile
- Electrical, electronics and telecommunications
- Pharmaceuticals
- Consumer durables
- Capital goods giants
- Banking and insurance
- Textile/garment

As such, any industrial entity today is in a position to call for latest technology that is available across the globe under any such terms and conditions that suits the organization.

2.5 IT in Competitive Advantage

Information technology has become the backbone of every business today. Investment in IT enabled projects has been increasing. However, IT is still not used much for strategic thinking. Information technology is not mere a computer application. Each organization is now aware of the effects, benefits and implication of information technology in business and also its ability in creating sustainable and maintainable competitive advantages. In business, information technology is used through the value chains of activities which, in turn, enable the organization to optimize and control function of operations for easy decision making. Strategic importance of IT lays its ability in developing the concept 'value chain' in organization by creating interdependent generic value activities comprising of elements such as suppliers, customers, production, finance, etc. IT creates a linkage between these activities through a value chain. It transforms the value chain activities and also the product's physical components and process either by lowering cost of value activities or by product's differentiation. IT also transforms the pace of competition itself.

Information technology can also add value to firm by reducing cost or increasing revenue. Information technology's ability in adding value to firm is different from its ability as being a source of sustainable competitiveness. However, the potential of information technology as a source of sustainable competitive advantage is in dispute and is based on certain assumptions. These assumptions have resulted in two emerging theories. The relationship between information technology and sustainable competitive advantage was initially studied by a theory called create-capture-keep paradigm. It was based on the switching cost of customers. This theory was replaced by resource based theory which explains the firm's ability to use IT to leverage the fundamental resources as advantage as a source of sustainable competitive advantage. Under this theory, when a firm possesses same capabilities or resources which competing firm does not possess, it constitutes a source of competitive advantage. However, if resource immobility incurs cost to the firm or brings disadvantages, it does not constitute sustainable competitive advantage. Sometimes, imitation of IT resources of successful firm without ambiguity or social complexity also brings sustainable competitive advantage. In any organization, information technology becomes potential sources of competitive advantages in 5 ways mentioned below:

1) when supplier becomes monopolist and IT is unique and essential for customers - switching cost of customers
2) when capital investment in IT enabled resources increases
3) when technology becomes as proprietary technology
4) when technical skills are needed to build IT application
5) when managerial skills are developed to understand and exploit IT application to increase business operations

Every information available in the organization can be effectively converted into fruitful result and can attain sustainable competitive advantage, if the following steps are observed:

- Assess existing and potential information intensity of product and process
- Evaluate the likely impact of IT on their industry structure
- Identify the rank of activities that information first affects
- Quantify how information technology brings new business
- Generate a plan of action for taking advantage of IT

Nowadays, IT is not at all a supporting system, but a way of thinking that contributes substantial competitive advantage having substantial effects and implications in the firm's operations. Competitive advantage comes when someone has or does something others can't, but as technology becomes more ubiquitous, it gets harder to extract that advantage from the IT investments and keep ahead of the chasing pack. The on-going economic situation and a more competitive environment including the potential of new competitors and pricing challenges, means traditional rules of competition will be challenged and companies will need to effectively compete and collaborate with others simultaneously by focusing on customer needs. Investing in individual IT innovations for competitive advantage can be a good thing, and if done properly can bring rewards, however, those rewards can be short lived or unsustainable unless the business case has been thought out properly. In fact, it can be argued that for competitive advantage to be sustainable, before looking at specific projects and their business cases, one needs to have things in order internally. In particular, this means a consistent technology platform (rather than a mismatch of legacy systems), consistent processes across the company with systems that are designed to support and improve those processes, automation of common processes and most importantly, ensuring those processes deliver value to the customer. Companies with the best processes will win in most or all markets. Those that deploy technology to improve their processes and succeed will see innovation spread rapidly. The success will prompt them to make even bolder and more frequent competitive moves and customers are likely to switch to them in response. Those companies will also be in a better position to strike back at competitors much more quickly and add further technology based innovation rather than just copy first to market opportunities.

Here are some thoughts for maximizing the competitive advantage in near future:

Sourcing Models: Some of the trends at the moment take the old outsourcing model and move it in a number of directions.

Cloud Computing: where you do not own the physical IT infrastructure and avoid capital expenditure by renting usage from a third party provider. The cloud type services include such things as data storage, computer servers and software as a service (SAAS). Though there may be an economic argument, however, the location and security of data remains a concern for the majority of companies where they hold sensitive data, certainly for the time being.

Shared Services Centre: It is normally the pooling of common services, repeated in a number of parts of the business into one operation, but one which has now gone outside the business into joint ventures. This can be difficult outside of the public sector to share common services with other companies including competitors, but if set up correctly, it could bring benefits to all parties with minimum risk.

Right Sourcing: It is a mixture of shared services, proactive management of vendors including virtual teams and using best of breed solutions. Here, one must account for government's initiative in implementing the concept IT techniques ultimately to facilitate the customer. For example,

- Railways - information, reservation, etc. > Online
- Income tax department - filing returns, information regarding tax assessment, circulars/notifications, etc. > Online
- Customs, central excise & service tax - filing returns, information regarding assessment, circulars/notifications, etc. > Online
- Sales tax/VAT - filing returns, information regarding tax assessment, circulars/notifications, etc.

2.6 Technology and Product Development

In business and engineering, new product development (NPD) is the term used to describe the complete process of bringing a new product to market. A product is a set of benefits offered for exchange and can be tangible (that is, something physical you can touch) or intangible (like a service, experience or belief). There are two parallel paths involved in the NPD process: one involves the idea generation, product design and detail engineering; the other involves market research and marketing analysis. Companies typically see new product development as the first stage in generating and commercializing new products within the overall strategic process of product life cycle management used to maintain or grow their market share.

Role of Private Sector

A strong, dynamic and sustainable basic research enterprise is a foundation for progress. For the goals of our society, i.e., a vigorous economy, strong national defense and a healthy populace and environment to be realized, a private sector capable of translating scientific discoveries into products, advances and other developments must be an active participant in the overall science and technology enterprise. For a technology-based company, the question is not really "Should the company do research?" but rather, "How much?" and "What kind of research should the organization focus on?" The needs of different companies vary greatly. Large and established companies often have greater resources available, but they may also have shareholders accustomed to regular dividends and unwilling to forgo them for the uncertain benefits of research that is more basic in nature, and therefore, more risky. Such a company may decide to stick to exploration that is largely aimed at refining its existing products or increasing production. For the small newcomer, pursuit of research, which is far riskier, may be the only way to break into a competitive market. Such young start-up companies must rely entirely on the initial capital provided by their investors to finance this research. Indeed, raising capital is the primary problem faced by many young companies. Some firms wrap up when the financial backing runs out with no product in sight. Others hit it big. A company's size, however, is not necessarily an indicator of what type of research it will do. Certain big and highly successful companies maintain research divisions whose purpose is to push the boundaries of their research and in doing so to risk more to ensure the company stays ahead of, and innovate faster than their competitors.

2.7 Competitive Advantage Through Technology

A generic environmental analysis pertaining to technological aspect helps us to achieve/create competitive advantage over the competitors in the business domain. Following important elements are to be accounted for:

A) Technology absorption, adoption, dispersal

Technology absorption: Most technology in 'latecomers' comes from abroad, in mixture of two forms:

- Embodied: in capital goods, patents, blueprints, designs, models, etc.
- Tacit: knowledge that can be 'transferred' only by close interaction and learning by new user.

Using technology efficiently, thus, needs conscious efforts by the enterprise and also the 'system' in which it works (suppliers, customers, technology support, training institutions, etc.). Technology absorption needs stable and conducive policy framework. Technology access is increasingly linked to foreign direct investment (FDI), but attracting, rooting and extracting benefits from FDI needs dynamic local firms and institutions.

Technology adoption: Technology adoption is important because it is the vehicle that allows most people to participate in a rapidly changing world where technology has become central to our lives. Individuals who won't or can't adopt will increasingly limit their ability to participate fully in the financial and convenience benefits associated with technology. Understanding the factors influencing technology adoption helps us predict and manage who adopts, when, and under what conditions. With this information, we can assess where people are in the adoption process and support them as they move from technology acceptance through to usage. Technology adoption process comprises of steps as under:

- Awareness – potential users learn enough about the technology and its benefits to decide whether they want to investigate further
- Assessment – potential users evaluate the usefulness and usability of the technology and the ease or difficulty of adopting
- Acceptance – potential users decide to acquire and use the technology or decide not to adopt
- Learning – users develop the skills and knowledge required to use the technology effectively
- Usage – users demonstrate appropriate and effective use of the technology

B) Evolution of IT/internet/wireless technology

This has witnessed into convenience, utility, punctuality and perfection within shortest time span, ultimately resulting in cost savings to great extent. Examples: banking (internet/phone/core), fund transfer, data transfer, e-commerce (buying/selling activity), communication (video-conferencing/tele-conferencing), etc.

C) Miniaturization of computing technologies

Miniaturization in computer technology refers to the evolution of electronic devices as they become smaller, faster and more efficient. The semiconductor industry is competitive and endlessly endeavors to replace yesterday's computer technology with smaller chips, with features as small as a thousandth of the width of a human hair. The idea is to get more transistors on a single chip. Increasing the number of transistors ultimately leads to more power in the computer, cell phone and other electronic devices that have a computer chip inside. Just about every electronic device has a computer chip inside. The pursuit to make these devices more powerful and, thus, effectively independent of a constant power source is the focus for miniaturization in computer technology. Ultimately, the goal is to have computers everywhere by making components smaller and more powerful. These days one can buy one megabyte of memory for less than a stick of gum.

Advantages of miniaturization

- Lower material costs
- Higher production output
- Potentially faster operation

Disadvantages of miniaturization

- Increased complexity in circuit design (may lead to poorer performance)
- Increased engineering/R&D costs
- Not serviceable after manufacture

D) Government's support in R&D

Governments generally offer/grant concessions to industries for carrying out R&D activity for upward growth in all aspects by means of calling for man, machine, material and time factors.

E) Major technological changes

- Conventional TVs v/s LCD, LED, 3D and plasma

- Mobile phones: nanotechnology > light weight/multi-sim/assorted functional features
- Computers: desktops v/s laptops
- Conventional concrete (manufacture and supply) v/s ready-mix concrete
- Factory fabricated v/s moulded/assembled furniture
- Internet: dial-up concept v/s broadband
- Internet: desktop v/s wi-fi
- Introduction of small cars: 800 (Maruti), Nano (Tata), Eon (Hyundai), Brio (Honda)
- Steel v/s engineering plastics/thermoplastics

2.8 Strategic Significance of Technology

The information revolution is sweeping through our economies. No company can escape its effects. Dramatic reductions in the cost of obtaining, processing and transmitting information are changing the way business is done. Most general managers know that the revolution is under way and few dispute its importance. As more and more of their time and investment capital is absorbed in information technology and its effects, executives have a growing awareness that the technology can no longer be the exclusive territory of EDP or IS departments. As they see their rivals use information for competitive advantage, these executives recognize the need to become directly involved in the management of the new technology. Managers must first understand that information technology is more than just computers. Today, information technology must be conceived broadly to encompass the information that businesses create and use as well as a wide spectrum of increasingly convergent and linked technologies that process the information. In addition to computers, the information technology infrastructure includes data recognition equipment, communications technologies, factory automation and other hardware and services. The information revolution is affecting competition in three vital ways:

- It changes industry structure and, in so doing, alters the rules of competition.
- It creates competitive advantage by giving companies new ways to outperform their rivals.
- It spawns whole new businesses, often from within a company's existing operations.

In the following, discussion is presented about the reasons why information technology has acquired strategic significance and how it is affecting all businesses. Subsequently, it is described how the new technology changes the nature of competition and how astute companies have exploited this. Finally, a procedure is outlined which managers can use to assess the role of information technology in their business and to help define investment priorities to turn the technology to their competitive advantage.

Information technology is changing the way companies operate. It is affecting the entire process by which companies create their products. Furthermore, it is re-shaping the product itself: the entire package of physical goods, services and information companies provide to create value for their buyers. An important concept that highlights the role of information technology in competition is the 'value chain'. This concept divides a company's activities into the technologically and economically distinct activities it performs to do business. These are called 'value activities'. The value a company creates is measured by the amount that buyers are willing to pay for a product or service. A business is profitable if the value it creates exceeds the cost of performing the value activities. To gain competitive advantage over its rivals, a company must either perform these activities at a lower cost or perform them in a way that leads to differentiation and a premium price. A company's value activities fall into nine generic categories. Primary activities are those involved in the physical creation of the product, its marketing and delivery to buyers and its support and servicing after sale. Support activities provide the inputs and infrastructure that allow the primary activities to take place. Every activity employs purchased inputs, human resources and a combination of technologies. Firm infrastructure, including such functions as general management, legal work and accounting, supports the entire chain. Within each of these generic categories, a company will perform a number of discrete activities, depending on the particular business. Service business, for example, frequently includes activities such as installation, repair, adjustment, upgrading and parts inventory management.

A company's value chain is a system of interdependent activities, which are connected by linkages. Linkages exist when the way in which one activity is performed affects the cost or effectiveness of other activities. Linkages often create trade-offs in performing different activities that should be optimized. This optimization may require trade-offs. For example, a costlier product design and more expensive raw materials can reduce after-sale service costs. A company must resolve such trade-offs, in accordance with its strategy, to achieve competitive advantage. Linkages also require activities to be coordinated. On-time delivery requires that operations, outbound logistics and service activities (installation, for example) should function smoothly together. Good coordination allows on-time delivery without the need for costly inventory. Careful management of linkages is often a powerful source of competitive advantage because of the difficulty rivals have in perceiving them and in resolving trade-offs across organizational lines.

The value chain for a company in a particular industry is embedded in a larger stream of activities that we term the 'value system'. The value system includes the value chains of suppliers, who provide inputs (such as raw materials, components and purchased services) to the company's value chain. The company's product often passes through its channels' value chains on its way to the ultimate buyer. Finally, the product becomes a purchased input to the value chains of its buyers, who use it to perform one or more buyer activities. Linkages not only connect value activities inside a company but also create interdependencies between its value chain and those of its suppliers and channels.

A company can create competitive advantage by optimizing or coordinating these links to the outside. For example, a candy manufacturer may save processing steps by persuading its suppliers to deliver chocolate in liquid form rather than in molded bars. Just-in-time deliveries by the supplier may have the same effect. However, the opportunities for savings through coordinating with suppliers and channels go far beyond logistics and order processing. The company, suppliers and channels can all benefit through better recognition and exploitation of such linkages.

Competitive advantage in either cost or differentiation is a function of a company's value chain. A company's cost position reflects the collective cost of performing all its value activities relative to rivals. Each value activity has cost drivers that determine the potential sources of a cost advantage. Similarly, a company's ability to differentiate itself reflects the contribution of each value activity toward fulfillment of buyer needs. Many of a company's activities, not just its physical product or service, contribute to differentiation. Buyer needs, in turn, depend not only on the impact of the company's product on the buyer but also on the company's other activities (for example, logistics or after-sale services). In the search for the competitive advantage, companies often differ in competitive scope or the breadth of their activities. Competitive scope has four key dimensions: segment scope, vertical scope (degree of vertical integration), geographic scope and industry scope (or the range of related industries in which the company competes). Competitive scope is a powerful tool for creating competitive advantage. Broad scope can allow the company to exploit interrelationships between the value chains serving different industry segments, geographic areas or related industries. For example, two business units may share one sales force to sell their products or the units may coordinate the procurement of common components. Competing nationally or globally with a coordinated strategy can yield a competitive advantage over local or domestic rivals. By employing a broad vertical scope, a company can exploit the potential benefits of performing more activities internally rather than use outside suppliers. By selecting a narrow scope, on the other hand, a company may be able to tailor the value chain to a particular target segment to achieve lower cost or differentiation. The competitive advantage of a narrow scope comes from customizing the value chain to best serve particular product varieties, buyers or geographic regions. If the target segment has unusual needs, broad-scope competitors will not serve it well.

A) Transforming the Value Chain

Information technology is permeating the value chain at every point, transforming the way value activities are performed and the nature of the linkages among them. It also is affecting competitive scope and reshaping the way products meet buyer needs. These basic effects explain why information technology has acquired strategic significance and is different from the many other technologies businesses use. Every value activity has both a physical and an information-processing component. The physical component includes all the physical tasks required to perform the activity.

The information-processing component encompasses the steps required to capture, manipulate and channel the data necessary to perform the activity. Every value activity creates and uses information of some kind. A logistics activity, for example, uses information like scheduling promises, transportation rates and production plans to ensure timely and cost-effective delivery. A service activity uses information about service requests to schedule calls and order parts and generates information on product failures that a company can use so as to revise product designs and manufacturing methods. An activity's physical and information-processing components may be simple or quite complex. Different activities require a different mix of the two components. For instance, metal stamping uses more physical processing than information processing or processing of insurance claims requires just the opposite balance. For most of industrial history, technological progress principally affected the physical component of what businesses do. During the industrial revolution, companies achieved competitive advantage by substituting machines for human labor. Information processing at that time was mostly the result of human effort.

Now the pace of technological change is reversed. Information technology is advancing faster than technologies for physical processing. The costs of information storage, manipulation and transmittal are falling rapidly and the boundaries of what is feasible in information processing are at the same time expanding. During the industrial revolution, the railroad cut the travel time from Boston, Massachusetts, to Concord, New Hampshire, from five days to four hours, a factor of 30. However, the advances in information technology are even greater. The cost of computer power relative to the cost of manual information processing is at least 8,000 times less expensive than the cost 30 years ago. Between 1958 and 1980, the time for one electronic operation fell by a factor of 80 million. Department of Defense studies show that the error rate in recording data through bar coding is 1 in 3,000,000, compared to 1 error in 300 manual data entries. This technological transformation is expanding the limits of what companies can do faster than managers can explore the opportunities. The information revolution affects all nine categories of value activity, from allowing computer-aided design in technology development to incorporating automation in warehouses. The new technology substitutes machines for human effort in information processing. Paper ledgers and rules of thumb have given way to computers.

Initially, companies used information technology mainly for accounting and record-keeping functions. In these applications, the computers automated repetitive clerical functions such as order processing. Today, information technology is spreading throughout the value chain and is performing optimization and control functions as well as more judgmental executive functions. General Electric, for instance, uses a database that includes the accumulated experience and (often intuitive) knowledge of its appliance service engineers to provide support to customers by phone. Information technology is generating more data as a company performs its diverse activities and is permitting it to collect or capture information that was not available before. Such technology also makes room for a more comprehensive analysis and use of the expanded data. The number of variables that a company can analyze or control has grown dramatically.

Hunt-Wesson, for example, developed a computer model to aid it in studying distribution-center expansion and relocation issues. The model enabled the company to evaluate many more different variables, scenarios and alternative strategies than had been possible before. Similarly, information technology helped Sulzer Brothers' engineers improve the design of diesel engines in ways that manual calculations could not. Information technology is also transforming the physical processing component of activities. Computer-controlled machine tools are faster, more accurate and flexible in manufacturing than the older and manually operated machines. Schlumberger has developed an electronic device permitting engineers to measure the angle of a drill bit, the temperature of a rock and other variables while drilling oil wells. The result: drilling time is reduced and some well-logging steps are eliminated. On the West Coast, some fishermen now use weather satellite data and ocean temperatures to identify promising fishing grounds. This practice greatly reduces the fishermen's steaming time and fuel costs.

Information technology not only affects how individual activities are performed through new information flows, it is also greatly enhancing a company's ability to exploit linkages between activities, both within and outside the company. The technology is creating new linkages between activities and companies can now coordinate their actions more closely with those of their buyers and suppliers. For example, McKesson, one of the largest drug distributors, provides its drugstore customers with terminals. The company makes it so easy for clients to order, receive and prepare invoices that the customers, in return, are willing to place larger orders. At the same time, McKesson has streamlined its order processing. Finally, the new technology has a powerful effect on competitive scope. Information systems allow companies to coordinate value activities in far-flung geographic locations (for example, Boeing engineers work on designs online with foreign suppliers). Information technology is also creating many new interrelationships among businesses, thus, expanding the scope of industries in which a company must compete to achieve competitive advantage. So pervasive is the impact of information technology that it confronts the managers and executives with a tough problem: too much information. This problem creates new uses of information technology to store and analyze the flood of information available to managers and executives.

B) Transforming the Product

Most products have always had both a physical and an information component. The latter, broadly defined, is everything that the buyer needs to know to obtain the product and use it to achieve the desired result. That is, a product includes information about its characteristics and how it should be used and supported. For example, convenient and accessible information on maintenance and service procedures is an important buyer criterion in consumer appliances. Historically, a product's physical component has been more important than its information component. The new technology, however, makes it feasible to supply far more information along with the physical product. For example, General Electric's appliance service data base supports a consumer hotline that helps differentiate GE's service support from its rivals'.

Similarly, some rail road and trucking companies offer up-to-the-minute information on the whereabouts of shippers' freight, which improves coordination between shippers and these companies. The new technology is also making it increasingly possible to offer products with no physical component at all. Compustat's customers have access to corporate financial data filed with the Securities and Exchange Commission and many companies have sprung up to perform energy use analyses of buildings. Many products also process information in their normal functioning. A dishwasher, for example, requires a control system that directs the various components of the unit through the washing cycle and displays the process to the user. The new information technology is enhancing product performance and is making it easier to boost a product's information content. Electronic control of the automobile, for example, is becoming more visible in dashboard displays, talking dashboards, diagnostic messages and the like.

There is an unmistakable trend toward expanding the information content in products. This component, combined with changes in companies' value chains, underscores the increasingly strategic role of information technology. There are no longer mature industries, rather, there are mature ways of doing business.

C) Direction and Pace of Change

Although a trend toward information intensity in the companies and products is evident, the role and importance of the technology differs in each industry. Banking and insurance, for example, have always been information intensive. Such industries were naturally among the first and most enthusiastic users of data processing. On the other hand, physical processing will continue to dominate in industries that produce, say, cement, despite increased information processing in such businesses. This relates information intensity in the value chain to information content in the product and illuminates the differences in the role and intensity of information among various industries. The banking and newspaper industries have a high information-technology content in both product and process. The oil-refining industry has a high use of information in the refining process, but a relatively low information content in the product dimension.

Due to the falling cost and growing capacity of the new technology, many industries seem to be moving toward a higher information content in both product and process. It should be emphasized that technology will continue to improve rapidly. The cost of hardware will continue to drop and managers will continue to distribute the technology among even the lower levels of the company. The cost of developing software, now a key constraint, will fall as more packages become available that are easily tailored to customers' circumstances. The applications of information technology that companies are using today are only a beginning. Information technology is not only transforming products and processes but also the nature of competition itself. Despite the growing use of information technology, industries will always differ in their position as well as their pace of change.

D) Changing the Nature of Competition

After surveying a wide range of industries, it is observed that information technology is changing the rules of competition in three ways. First, advances in information technology are changing the industry structure. Second, information technology is an increasingly important lever that companies can use to create competitive advantage. A company's search for competitive advantage through information technology often also spreads to affect industry structure as competitors imitate the leader's strategic innovations. Finally, the information revolution is spawning completely new businesses. These three effects are critical for understanding the impact of information technology on a particular industry and for formulating effective strategic responses.

E) Changing Industry Structure

The structure of an industry is embodied in five competitive forces that collectively determine industry profitability: the power of buyers, the power of suppliers, the threat of new entrants, the threat of substitute products and the rivalry among existing competitors. The collective strength of the five forces varies from industry to industry, as does average profitability. The strength of each of the five forces can also change, either improving or eroding the attractiveness of an industry. Information technology can alter each of the five competitive forces and, hence, industry attractiveness as well. The technology is unfreezing the structure of many industries, creating the need and opportunity for change. For example:

- Information technology increases the power of buyers in industries assembling purchased components. Automated bills for materials and vendor quotation files make it easier for buyers to evaluate sources of materials and make-or-buy decisions.
- Information technologies requiring large investments in complex software have raised the barriers to entry. For example, banks competing in cash management services for corporate clients now need advanced software to give customers online account information. These banks may also need to invest in improved computer hardware and other facilities.
- Flexible computer-aided design and manufacturing systems have influenced the threat of substitution in many industries by making it quicker, easier and cheaper to incorporate enhanced features into products.
- The automation of order processing and customer billing has increased rivalry in many distribution industries. The new technology raises fixed costs at the same time as it displaces people. As a result, distributors must often fight harder for incremental volume.
- Industries such as airlines, financial services, distribution and information suppliers have felt these effects so far.

Information technology has had a particularly strong impact on bargaining relationships between suppliers and buyers since it affects the linkages between companies and their suppliers, channels and buyers. Information systems that cross company lines are becoming increasingly common these days. In few cases, the boundaries of industries themselves have changed. Systems that connect buyers and suppliers are spreading. Xerox gives manufacturing data to suppliers electronically to help them deliver materials. To speed up order entry, Westinghouse Electric Supply Company and American Hospital Supply have furnished their customers with terminals. Among other things, many systems raise the costs of switching to a new partner because of the disruption and re-training required. These systems tend to tie companies more closely to their buyers and suppliers. Information technology is altering the relationship among scale, automation and flexibility with potentially profound consequences. Large-scale production is no longer essential to achieve automation. As a result, entry barriers in a number of industries are falling.

At the same time, automation no longer necessarily leads to inflexibility. For example, General Electric rebuilt its Erie locomotive facility as a large-scale yet flexible factory using computers to store all design and manufacturing data. Ten types of motor frames can be accommodated without manual adjustments to the machines. After installation of a 'smart' manufacturing system, BMW can build customized cars (each with its own tailored gearbox, transmission system, interior and other features) on the normal assembly line. Automation and flexibility are achieved simultaneously, a pairing that changes the pattern of rivalry among competitors. The increasing flexibility in performing many value activities combined with the falling costs of designing products has triggered an avalanche of opportunities to customize and to serve small market niches. Computer-aided design capability not only reduces the cost of designing new products, but also dramatically reduces the cost of modifying or adding features to existing products. The cost of tailoring products to market segments is falling, again affecting the pattern of industry rivalry.

Though managers can use information technology to improve their industry structure, however, the technology also has the potential to destroy this structure. For example, information systems now permit the airline industry to alter fares frequently and to charge many different fares between any two points. At the same time, however, the technology makes the flight and fare schedules more readily available and allows travel agents and individuals to shop around quickly for the lowest fare. The result is a lower fare structure than might otherwise exist. Information technology has made a number of professional service industries less attractive by reducing personal interaction and making service more of a commodity. Therefore, the managers must look carefully at the structural implications of the new technology to realize its advantages or to be prepared for its consequences.

F) Creating Competitive Advantage

In any company, information technology has a powerful effect on competitive advantage in either cost or differentiation. The technology affects value activities themselves or allows companies to gain competitive advantage by exploiting changes in competitive scope. As has been seen, information technology can alter company's costs in any part of the value chain. The technology's historical impact on cost was confined to activities in which repetitive information processing played a large part. These limits no longer exist, however. Even activities like assembly that mainly involve physical processing now have a large information-processing component. Canon, for example, built a low-cost copier assembly process around an automated parts-selection and materials-handling system. Assembly workers have bins containing all the parts needed for the particular copier. Canon's success with this system derives from the software that controls parts inventory and selection. In insurance brokerage, a number of insurance companies usually participate in underwriting a contract. The costs of documenting each company's participation are high. Now a computer model can optimize (and often reduce) the number of insurers per contract, thus, lowering the broker's total cost. In garment production, equipment such as automated pattern drawers, fabric cutters and systems for delivering cloth to the final sewing station have reduced the labor time for manufacturing by up to 50%.

In addition to playing a direct role in cost, information technology often alters the cost drivers of activities in ways that can improve (or erode) a company's relative cost position. For example, Louisiana Oil & Tire has taken all ten of its salespeople off the road and made them into telemarketers. As a result, sales expenses have fallen by 10% and sales volume has doubled. However, the move has made the national scale of operations the key determinant of the cost of selling, rather than regional scale.

- **Enhancing differentiation:** The impact of information technology on differentiation strategies is equally dramatic. As noted earlier, the role of a company and its product in the buyer's value chain is the key determinant of differentiation. The new information technology makes it possible to customize products. Using automation, for instance, Sulzer Brothers has increased from five to eight the number of cylinder bore sizes of new low-speed marine diesel engines. Ship-owners now choose an engine that is more precisely suited to their needs and thereby recoup significant fuel savings. Similarly, Digital Equipment's artificial intelligence system, XCON, uses decision rules to develop custom computer configurations. This dramatically reduces the time required to fill orders and increases accuracy which enhances Digital Equipment's image as a quality provider. By bundling more information with the physical product package sold to the buyer, the new technology affects a company's ability to differentiate itself.

- **Changing competitive scope:** Information technology can alter the relationship between competitive scope and competitive advantage. The technology increases a company's ability to coordinate its activities regionally, nationally and globally. It can unlock the power of broader geographic scope to create competitive advantage. Consider the newspaper industry. Dow Jones, publisher of the Wall Street Journal, pioneered the page transmission technology that links its 17 U.S. printing plants to produce a truly national newspaper. Such advances in communication plants have also made it possible to move toward a global strategy.

The information revolution is creating interrelationships among industries that were previously separate. The merging of computer and telecommunications technologies is an important example. This convergence has profound effects on the structure of both industries. For example, AT&T is using its position in telecommunications as a staging point for entry into the computer industry. IBM, which recently acquired Rolm, the telecommunications equipment manufacturer, is now joining the competition from the other direction. Information technology is also at the core of growing interrelationships in financial services, where the banking, insurance and brokerage industries are merging as well as in office equipment, where once distinct functions such as typing, photocopying and data and voice communications can now be combined. Broad-line companies are increasingly able to segment their offerings in ways that were previously feasible only for focused companies. In the past, each local office set prices using manual procedures. Intermodal now uses microcomputers to link its offices to a center that calculates all prices. The new system gives the company the capacity to introduce a new pricing policy to offer discounts to national accounts, which place their orders from all over the country. As information technology becomes more widespread, the opportunities to take advantage of a new competitive scope will only increase. The benefits of scope, however, can accrue only when the information technology spread throughout the organization can communicate. Completely decentralized organizational design and application of information technology will thwart these possibilities because the information technology introduced in various parts of a company will not be compatible.

G) Spawning New Businesses

The information revolution is giving birth to completely new industries in three distinct ways. First, it makes new businesses technologically feasible. For example, modern imaging and telecommunications technologies blend to support new facsimile services such as Federal Express's Zapmail. Similarly, advances in microelectronics made personal computing possible. Services such as Merrill Lynch's Cash Management Account required new information technology to combine several financial products into one. Second, information technology can also spawn new businesses by creating derived demand for new products. One example is Western Union's EasyLink service, a sophisticated, high-speed and data-communications network that allows personal computers, word processors and other electronic devices to send messages to each other and to telex machines throughout the world. This service was not needed before the spread of information technology caused a demand for it. Third, information technology creates new businesses within old ones. A company with information processing embedded in its value chain may have excess capacity or skills that can be sold outside. Sears took advantage of its skills in processing credit card accounts and of its massive scale to provide similar services to others. It sells credit-authorization and transaction-processing services to Phillips Petroleum and retail remittance-processing services to Mellon Bank. Similarly, a manufacturer of automotive parts, A.O. Smith, developed data-communications expertise to meet the needs of its traditional businesses. As a bank consortium went looking for a contractor to run a network of automated teller machines, A.O. Smith got the job. Eastman Kodak recently began offering long-distance telephone and data-transmission services through its internal telecommunications system. If the information technology used in a company's value chain is sensitive to scale, a company may improve its overall competitive advantage by increasing the scale of information processing and lowering costs. By selling extra capacity outside, it is at the same time generating new revenue.

Companies also are increasingly able to create and sell to others information that is a by-product of their operations. National Benefit Life reportedly merged with American Can in part to gain access to data on the nine million customers of American Can's direct-mail retailing subsidiary. The use of bar-code scanners in supermarket retailing has turned grocery stores into market research labs. Retailers can run an ad in the morning newspaper and find out its effect by early afternoon. They can also sell this data to market research companies and to food processors.

H) Competing in the Age of Information

Senior executives can follow the following steps to take advantage of opportunities that the information revolution has created.

1. Assess information intensity. A company's first task is to evaluate the existing and potential information intensity of the products and processes of its business units. To help managers accomplish this, some measures of potential importance of information technology need to be developed.
2. It is very likely that information technology will play a strategic role in an industry that is characterized by one or more of the following features:
    a. Potentially high information intensity in the value chain - a large number of suppliers or customers with whom the company deals directly, a product requiring a large quantity of information in selling, a product line with many distinct product varieties, a product composed of many parts, a large number of steps in a company's manufacturing process, a long cycle time from the initial order to the delivered product, etc.
    b. Potentially high information intensity in the product - a product that mainly provides information, a product whose operation involves substantial information processing, a product whose use requires the buyer to process a lot of information, a product requiring especially high costs for buyer training, a product that has

many alternative uses or is sold to a buyer with high information intensity in his or her own business, etc.

c. These may help identify priority business units for investment in information technology. If selecting priority areas, it is necessary to remember the breadth of information technology as it involves more than simple computing.

3. Determine the role of information technology in industry structure. Managers should predict the likely impact of information technology on their industry's structure. They must examine how information technology might affect each of the five competitive forces. Not only is each force likely to change, but industry boundaries may change as well. Chances are that a new definition of the industry may be necessary. Many companies are partly in control of the nature and pace of change in the industry structure. Companies have permanently altered the bases of competition in their favor in many industries through aggressive investments in information technology and have forced other companies to follow. Citibank, with its automated teller machines and transaction processing; American Airlines, with its computerized reservations system; and USA Today, with its newspaper page transmission to decentralized printing plants, are pioneers that have used information technology to alter industry structure. A company should understand how structural change is forcing it to respond and look for ways to lead change in the industry.
4. Identify and rank the ways in which information technology might create competitive advantage. The starting assumption must be that the technology is likely to affect every activity in the value chain. Equally important is the possibility that new linkages among activities are being made possible. By taking a careful look, managers can identify the value activities that are likely to be most affected in terms of cost and differentiation. Obviously, activities that represent a large proportion of cost or that are critical to differentiation bear closest scrutiny, particularly if they have a significant information-processing component. Activities with important links to other activities inside and outside the company are also critical. Managers and executives must examine such activities for ways in which information technology can create sustainable competitive advantage.

In addition to taking a hard look at its value chain, a company should consider how information technology might allow a change in competitive scope. Can information technology help the company serve new segments? Will the flexibility of information technology allow broad-line competitors to invade areas that were once the province of niche competitors? Will information technology provide the leverage to expand the business globally? Can managers harness information technology to exploit interrelationships with other industries? Or, can the technology help a company create competitive advantage by narrowing its scope? Also:

- A fresh look at the company's product may also be in order
- Can the company bundle more information with the product?
- Can the company embed information technology in it?

Mangers should investigate how information technology might spawn new businesses. Managers should also consider opportunities to create new businesses from existing ones. Information technology is an increasingly important avenue for corporate diversification. Lockheed, for example, entered the database business by perceiving an opportunity to use its spare computer capacity. Also needed to consider are:

- Identifying opportunities to spawn new businesses requires answering questions such as:
    - What information generated (or potentially generated) in the business could the company sell?
    - What information-processing capacity exists internally to start a new business?
    - Does information technology make it feasible to produce new items related to the company's product?

Managers should develop a plan for taking advantage of information technology. The first four steps should lead to an action plan to capitalize on the information revolution. This action plan should rank the strategic investments necessary in hardware and software as well as in new product development activities that reflect the increasing information content in products. Organizational changes that reflect the role that the technology plays in linking activities inside and outside the company are likely to be necessary. The management of information technology can no longer be the sole province of the EDP department. Increasingly, companies must employ information technology with a sophisticated understanding of the requirements for competitive advantage. Organizations need to distribute the responsibility for systems development more widely in the organization. At the same time, general managers must be involved to ensure that cross-functional linkages, more possible to achieve with information technology, are exploited.

These changes do not mean that a central information technology function should play an insignificant role. Rather than to control information technology, however, an IS manager should coordinate the architecture and standards of the many applications throughout the organization, as well as provide assistance and coaching in systems development. Unless the numerous applications of information technology inside a company are compatible with each other, many benefits may be lost. Information technology can help in the strategy implementation process. Reporting systems can track progress toward milestones and success factors. By using information systems, companies can measure their activities more precisely and help motivate managers to implement strategies successfully.

3

Strategy of Technology Management

3.1 Introduction

In the previous units, the meaning of technology was stated and the significance of technology management was explained. Technology adds value to an organization by providing better products, services and more effective processes. Rapid technological growth and globalization have made technological innovation vital to the survival and growth of businesses. Organizations cannot afford to ignore a systematic and planned approach to manage technology. Technology provides a path to organization's growth in three areas:

1. Differentiate products and develop new products for customer satisfaction
2. Develop new business opportunities
3. Reduce costs of its operations facilitating higher sales and profits

Management of organizations of high growth countries like Japan, Europe or the United States have long back recognized that the development area of new products and processes (innovation) offers greatest opportunities for growth of organizations. Hence, technology becomes a major source of competitive advantage. The technology was also studied in the earlier unit in context of competitive advantages. As it is known, one of the important keys to competitive success is to develop a system for identifying areas of technological leadership. At the same time, the company needs to be alert for areas where competitors try to gain its business. Hence, a strategic approach to management of technology is necessary. Related to the issue of managing technology is 'technology life cycle'. Understanding of technology life cycle and its stages helps in evolving an effective technology strategy. One of the most important components of managing technology is research and development (R&D) process. Managing research and development is of vital importance to many sectors of businesses. Research and development management has become complex. Identifying issues involved and trends in management of research and development are helpful in context of developing technology strategy.

In view of the fact that technology is imperative and inescapable, most of the decisions related to technology have become strategic. Strategic decisions on technology have the following features:

- The decisions are taken by top management
- They have long term perspective
- Risk involved is high
- Investment required also is high most of the time

3.2 Technology Strategy

The technology strategy is represented by business decisions and actions that guide the organization in search of success and technological innovation. The strategy is to be formulated and implemented in a business environment which is global, competitive and increasingly complex. Technology strategy plays a fundamental role in the life of technology-based companies. Technology strategy has become much more important and, at the same time, much more difficult to set and implement.

Technology strategy can be stated as the process of building, maintaining and exploiting a company's technological assets. Technology has become one of the basis of strategic planning. For competitiveness and survival of the businesses in today's environment, the management needs to find appropriate answers to complex questions, some of them are stated below:

- what future you want for the company?
- which technological strategy one has to follow to ensure this future?
- how to create and develop a climate which is favorable to innovation?
- which level of resources to allocate and how to develop this activity inside the corporation and in the market?

The objective of technology strategy is to gain competitive advantage by guiding an organization in the following:

- Getting/acquiring technology
- Developing technology

- Implementing/applying technology

As studied in the previous unit, competitive strategy can lead to two types of competitive advantage:

- Lower cost
- Differentiation

Depending on the scope of the organization's target market within its industry, these two types of competitive advantage lead to three generic competitive strategies:

A) Cost leadership: The organization seeks industry-wide cost leadership in serving a broad range of industry segments.

B) Differentiation: The organization seeks industry-wide differentiation in serving a broad range of industry segments.

C) Focus: The organization directs its entire strategy at a narrow target business segment, foregoing sales to other segments.

The decisions linked to technology strategy can be broadly categorized in three groups as shown below (Figure 3.1).

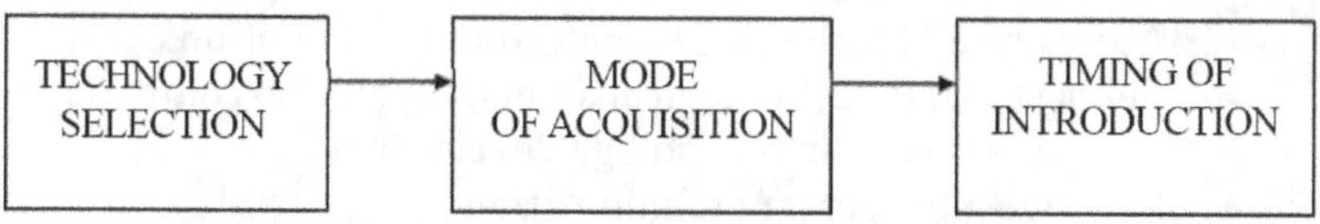

Figure 3.1 Decisions linked to technology strategy.

A) Technology Selection: These are the decisions on selection of technology/ies which needs to be acquired. This is based on identification of technologies necessarily required to meet company's objectives and checking whether such technology capability is available with the organization. In case such skills/capability is not available, a decision is made to acquire the technologies.

B) Mode of Acquisition: These are decisions on choosing among several alternatives and modes for acquisitions of technology. Several alternatives like developing technology in-house, seeking collaboration, joint venture and buying a company specializing in the technology need to be considered and carefully weighed for their advantages and disadvantages from strategic perspective.

C) Timing of Introduction: Timing decision is to decide at what time to introduce the new product. This requires consideration of several factors. Launching time plays a significant role in the success of a high-tech product.

3.2.1 Tasks in Technology Strategy

Technology strategy is concerned with the following tasks/processes:

- Identification of the critical technologies on which the company's competitiveness depends
- Identification and providing for distinctive technological competencies are necessary for establishing and/or retaining competitive advantage
- Feasibility of identified technologies and availability
- Identification of methods of acquisition for those technologies
- Building technology into product design
- Method of evaluation of these technologies
- Acquisition of technology amongst different options
- Get best ROI in technological assets/capability/competency
- Investment level in technology development
- Effective management of innovation
- Effective management of research and development function and projects
- Introduction of technology to the market

3.3 Models of Technology Strategy

Several models for technology strategy have been evolved. Some of the popular/known models are listed below:

A) Porter Model (1985): Porter's model of value chain is one of the best known and widely applied models of a company's value-creation processes. According to Porter, "competitive advantage cannot be understood by looking at a firm as a whole. It stems from the many discrete activities a firm performs in designing, producing, marketing, delivering and supporting its product. Each of these activities can contribute to a firm's relative cost position and create a basis for differentiation."

B) Hax & Majluf's Model (2001): It is contended that strategy formulation is one of the two major cycles in strategic planning that are intended to frame the key strategic issues of a firm through a sequential involvement of corporate, business and functional perspectives. The strategy formulation process would affect the second cycle of strategic and operational budgeting that deals with the final definition and subsequent consolidation at corporate level of the budgets for all the businesses and functions of the firm. The budget constitutes the legitimate output of this process, since it represents the commitments for strategy implementation.

C) A.D. Little's Model (1981): Little's model of technology management involves identification of the technologies, selection of the technology to achieve key factors of success, determination of the firm's strengths and weaknesses and technology strategy formulation. A.D. Little's model consists of two important dimensions, the competitive position and industry maturity.

3.4 Information Technology Strategy

Effective use of IT (information technology) is of crucial importance for managing organizations. Application of IT in organization requires significant investment, takes long time and requires skilled personnel to run the system. To get maximum benefits from IT, a holistic approach keeping in mind likely changes in internal and external environment of business and also the changes in IT field itself needs consideration. IT strategy helps in maximizing the return on IT investments and creates value. Information technology strategy requires consideration of organization wide processes. Information technology strategy requires developing a framework within which business applications can be deployed identifying necessary software, hardware and other infrastructure. Some of the features of IT strategy are:

- It is a process not a point in time event

- It is iterative - success comes after multiple - do and learn - cycles
- The key is the alignment of business and IT capability rather than designing IT to address business requirements

IT strategy sets direction for IT function in an organization as it:

- Ensures that maximum amount of money spent on value creation activities for the business
- Ensures that these funds create the maximum value

One must know the following properly for developing effective information technology strategy:

- Thorough understanding of business strategy of the company
- Deep insight and understanding of how technology will add value to business processes
- Clarity on principles to be followed on technology use

A) Business Strategy: A thorough understanding of business strategy is of vital importance. Ultimately, the information technology strategy should help in realizing objectives of business strategy. If information technology strategy is not in line with business strategy, organization may fail in meeting its objective. Due to strength and potential of information technology, success of business strategy is more dependent on information technology strategy.

B) Business Value: Information technology can be used in all the processes of an organization. Use of information technology in any process always results in some value addition. This value addition may be in the form of improving process or reducing time or gaining competitive advantage. For evolving information technology strategy, a detailed study and analysis is needed to identify processes for value addition and how this is achieved.

C) Technology Use: In developing an information technology strategy, it is helpful in clarifying some basic principles in the use of technology. These principles should be easily understood by all the concerned employees. Some of the commonly used principles are explained below.

D) Consistency: Data should be entered only once. The same data should be available to all concerned. These principles ensure that information received or sent through documents/reports is consistent across the company. Simply stated, the principle is: enter once, use everywhere.

- **Reliability**: One should always verify, check the data to ensure its reliability. The process for changes in data should be as less as possible. The data should be what customer/user needs. The principle is: measure twice, cut once.
- **Maintainability**: The system should be easy to maintain. Complexities in the system should be avoided. This is easy to state and hardest to implement. Devoting significant amount of resources to maintain systems is undesirable. The principle is: keep it simple.

It can be noticed that there is no reference to hardware and software. Hardware and software are organization specific, hence, no generalization is possible. Alignment of technology strategy with business strategy is most important, as only then information technology strategy will effectively be able to add value to the business.

3.5 Technology Life Cycle Management

For developing technology strategy, we are interested in knowing whether the technology is worth an investment. In future, how will the technology develop? Will technology grow or will it decline in the future? Technology life cycle stages help in getting answer to these questions. There are several curves/models developed for technology life cycle. One of the well-known curves called 'S curve' is shown in Figure 3.2.

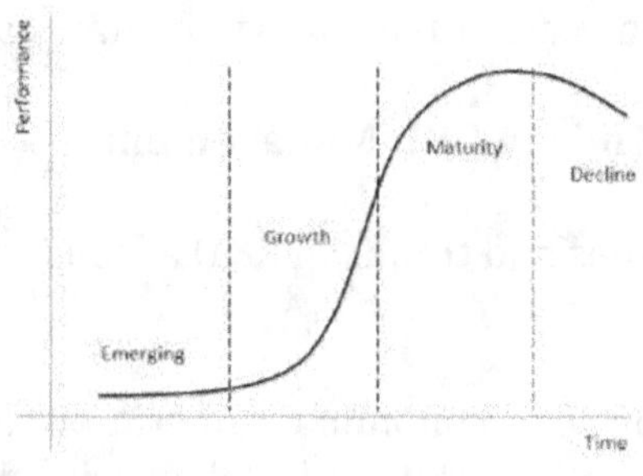

Figure 3.2 Technology life cycle.

In S Curve, x axis indicates time period and y axis represents performance. The S curve has four stages:

- Emergence
- Growth
- Maturity
- Decline

A) Emergence Stage: At emerging stage, new technology is conceived and introduced. At this stage, emergence of technology in context of intellectual property is considered. For example, in manufacturing industry, earlier steam operated machines and equipment were used. Afterwards, electrical motor technology emerged.

B) Growth Stage: During growth stage, technologies grow very fast and a dominant design is accepted. This dominant design depends largely on market acceptance of the technology. Customers can find the technology that meets their requirement or a solution for a problem they face. For example, in automobile industry, diesel/petrol based engine became a dominant design. Subsequently, the industry focused its efforts in development and improvement in the engine technology.

C) Maturity Stage: The third stage is called maturity stage. At this stage, technology is at its peak. Everyone is involved and customer, competitors as well as suppliers understand the technology. Major players get established. There is little scope for a new entrant. Revenues start indicating a declining trend. Technology may create opportunities for entirely new technology application.

D) Decline Stage: The fourth stage is the declining phase. At this stage, technology reaches a plateau. Sometimes, a new technology virtually replaces the earlier technology or changes its functionality.

3.6 Issues and Trends in R&D Management

There are many issues and trends in R&D as follows:

A) Issues in R&D

OECD (Organization for Economic Co-operation and Development) defines R&D as, "creative work undertaken on a systematic basis in order to increase the stock of knowledge, including knowledge of man, culture and society as well as the use of this stock of knowledge to devise new applications."

We understand research and development as something dealing with developing particular technology or some scientific work. It is considered to be carried out in laboratories, research centers or institutions by government and business organizations. One of the major functions of successful R&D set-ups is to generate innovative ideas. The R&D organizations convert these innovative ideas/concepts into products and services for economic gain, through the organizational system. Though innovation is not a random process, however, managers often argue that R&D performance is hard to measure and even more difficult to manage. An exploratory field study into technology-oriented R&D environments determines the principle factors that influence innovation-based performance of R&D teams. The results identify specific barriers and drivers to innovative team performance and provide insight into the type of an organizational environment and managerial leadership that is conducive to innovative R&D team performance. The data further suggests that many of the performance variables have their locus outside the R&D organization. Yet, managerial leadership style, both at the R&D team level and at senior management, has significant impact on creativity that ultimately affects R&D performance. Industry and government spend large amount of money for developing new or better products and services. These spending are called as R&D costs by accountants. The issue involved is about treating these outlays as assets or expenditure. A guideline is available based on the period of benefits from any outlays. If the outlay provides economic benefit for less than a year, then it is classified as an expense. If the benefit will be for more than one year, it is considered as an asset.

Let us take an example of pharmaceutical industry. It takes several years of research and large R&D expenditure to develop a drug. The drug process/product technology is patented and till another superior product is launched successfully by competitors, the company harvests significant profits for years to come. The accountants treat the research and development outlays not as an asset, but as expenditure. One of the reasons being the harsh truth: very few hi-tech products are commercially successful. Further, many research and development projects develop new products/processes that have to be abandoned or postponed due to technical constraints or enormous cost/time overruns. Another issue is about difficulty in assessing useful life. If research and development cost is to be considered as an asset, for how many years this cost should be amortized/spread. To assess life cycle of hi-tech product is difficult due to complexity. Several factors like competitor's strategy, technology and change in buyer's preference influence life cycle of such product. In the current business environment, most important objective of an organization is to maximize returns to shareholders. So, the focus of the top management is on short term profitability.

Research and development efforts facilitate long term growth and profitability at high initial cost. Hence, there is a tendency among CEOs to avoid large investment in R&D, as it adversely affects short term profitability. However, many organizations have recognized need for effective research and development for facing global competition by necessary strategies to introduce/improve their products and processes. Accordingly, ptogressive investments in R&D activities have been made. Attention on R&D has been focused with typical illustrative objectives:

- To gain technology leadership; to develop and sustain innovation
- To enhance quality and image
- To develop new product/processes for customer satisfaction
- To become market leader or to have increase market share in global/local market
- To ward off competition and to reduce cost/process time

Tremendous scope exists in improving effectiveness of R&D in developing countries. One of the weak areas of R&D management practices is effective project monitoring and control. Some of the observations on research and development organization practices are given below:

- Ad-hoc systems for R&D project management
- Lack of use of effective formal and structured effective planning and control techniques
- Feedback systems are reactive (rather than proactive)
- Cost is the most important criterion to assess performance of project for government
- Research and development organization
- Time is considered most important for private research and development organizations

R&D projects are characterized by high risk and uncertainty. The inherent characteristics of R&D projects make the planning effort difficult. These characteristics coupled with the requirement to reduce the cycle time make the R&D project monitoring and control process critical.

B) Trends in R&D

The rapid change in technologies, globalization of markets and the regionalization of technical expertise are forcing companies to give utmost importance to R&D. Bigger technology companies have to adjust their R&D organization due to these factors. Following are the trends in research and development:

1) Globalization of research and development: Large number of R&D laboratories and centers have been set up in different countries. As an example, in India, at Bangalore itself there are over 100 R&D laboratories set up in the last decade. Companies like Dow Chemicals have research laboratories in many developed and developing countries.

2) Increase in investment in research and development: Increased spending by most of the technology based companies is observed. At national levels too, expenditure on research and development is showing an increasing trend. For instance, EU has set a research and development expense target of 3%. In the year 2005, it was less than 2.5%.

3) Increased research and development in service sector (other than IT): A noticeable increase in research and development investment and efforts in service sector like health care and hospitality is seen. Growing competition and demanding customers are necessitating research and development efforts to develop new products/services to satisfy customer's needs.

4) Attention by top management: Top management of the companies are now sparing more time and attention on research and development matters. This change in trend is due to the important role of research and development in adding value and growth of the company.

5) Focus on applied research: Lesser importance is being given to pure research areas. The thrust area is on applied research aimed at developing product/process of commercial importance. The thrust is on result rather than efforts/process or knowledge gained.

6) Better monitoring and control: Research and development project monitoring and control has become more effective partly due to the use of IT tools/systems. The need for effective project monitoring has arisen due to the importance of timely/early completion of research and development projects.

7) Outsourcing research and development processes: There is a growing trend for outsourcing research and development processes. Four main reasons for the growing trend of outsourcing are:

- Cost reduction
- Help in timely completion of project
- Non-core processes
- Availability of expertise

8) Strategic alliances/collaborative effort: Many organizations do not have their own established research laboratory with infrastructure for all the technologies and processes required in the present and future. Setting up a research laboratory requires a long time and significant investment. All organizations may not have expertise in all the technologies. Hence, the trend is to have alliances/collaboration among organizations for specific research projects. It can be a development of technology, a product or processes.

4

Technology Forecasting

4.1 Introduction

The necessity to have a strategic view on technology related decisions was learnt in the previous units. These strategic decision areas include process technology and product technology and their continuing improvement, technology strategy, research and development (R&D), technology policy, technology project management, technology transfer, technology assessment, technological forecasting and innovation. Among the above, technology forecasting is an area which is linked directly to almost all major aspects of management of technology. As a matter of fact, technology forecasts are embedded in all strategic decisions related to technology in an organization and technologies influence all strategic business decisions. Louis V. Gerstner Jr., former chairman of the board and chief executive officer of IBM credited with turning around IBM's fortunes, in an interview in Business Week stated that the most troubling challenge facing business executives is that they must make plans for a decade on the basis of technologies that are changing every year. Organizations engaged in high-tech products or services have to necessarily be fully aware of technologies of future. If needed, organizations have to carry out technology forecast study for making long term decisions. Other organizations too need to keep updated on latest technology trends in processes and products.

Forecasting in simple terms means predicting future. Forecasting can be defined as reasonably definite statement about the future. It is a method for extrapolating the future. The extrapolation is done from the analysis of the current and past data. Forecasting is a process of predicting the future based on the past and present data. In forecasting, analysis of current trends is carried out and predictions are made about future based on those trends. In forecasting, we make statement about future events whose occurrence and outcome is not observed and is probabilistic. Forecasting is also viewed as a planning tool that helps management in making decisions coping uncertainties and risks of future. Forecasting provides information about the potential future events and their consequences for the organization. Forecasting does not reduce any risk or uncertainty about future, but it helps the management to consider these risks and uncertainties and make better decisions meeting its objectives. The purpose of any type of forecasting is to assist the decision-makers in choosing the policies and making promising.

4.2 Theory of Technology Forecasting

If professionals like technocrats, operations executives, management consultants and designers are asked about the definition of technology forecasting, completely different answers are expected. Technology forecasting has different perceptions. A technological forecast is concerned with the features/characteristics of the technology. A common characteristic is the level of technical performance like the capacity of a cargo ship, accuracy of hitting the target by a pilot-less fighter plan, etc. The forecast does not tell us how to achieve these characteristics. B. R. Martin's definition of technology forecasting is one of the most comprehensive and widely quoted: Technology forecasting is the process involved in systematically attempting to look into the longer-term future of science, technology, economy and society with the aim of identifying the areas of strategic research and emerging generic technologies likely to yield the greatest economic and social benefits. Technology forecasting involves a systematic, quantitative as well as qualitative analysis of technology. Technology forecasting includes the following:

- Project technological capabilities
- Predict the invention and
- Spread of technological innovations

Numerous techniques for technology forecasting were evolved in the 1960s in both business and defense applications. Nowadays, a term 'foresight studies' is increasingly used for technology forecasting. Figure 4.1 provides a generic process of technology forecasting indicating major inputs and outputs.

There are several assumptions about forecasting:

1. With complete certainty, we cannot state what the future will be.
2. There will always be an element of uncertainty in outcome/forecast regardless of the methods that we use.

3. There will always be blind spots in forecasts.
4. Providing forecasts to policy-makers will help them formulate social policy. The new social policy, in turn, will affect the future, thus, changing the accuracy of the forecast.

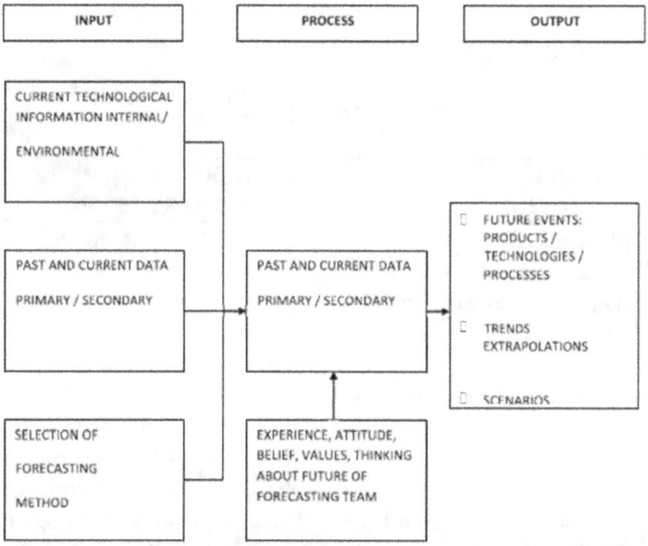

Figure 4.1 Technology forecasting process.

4.2.1 Uses of Technology Forecast

Major uses of technology forecasts at macro (country) level and micro (organization) level are stated in Table 4.1.

4.2.2 Technology Forecast and R&D

Technology forecasting has the following uses for research and development and product design functions:

- Output of technology forecast provides direction for taking up new research and development projects or new product for development
- Facilitates review of existing research and development, new product development projects for their continuation, modification or abandonment

- Identifies new performance level/parameters for new technology/product or process
- Helps in identifying types of resources-skills/capability, equipment, etc., required for effective management of research and development projects
- Indicates direction of investment required in research and development efforts

Table 4.1 Uses of technology forecasting

MICRO LEVEL (ORGANIZATION LEVEL)	MACRO LEVEL (COUNTRY LEVEL)
Provide inputs for new product/process development and/or improvement of existing products/processes. Set benchmarks (performance standards) for new products and processes Set technical parameters for new products/processes Help in strategic decisions by providing information on opportunities, challenges and threats in technology environment Facilitate strategic decision for time to launch new technology/products/processes	Provide inputs for developing a long term technology policy for the country. Help in identifying core areas for R&D, investments in technology Help in identifying thrust areas for technology: planning for sustaining/creating competitive advantages in technology to meet global challenges Planning for use of technology for country's strategic aims and growth, development and wellbeing of its citizens.

4.3 Methods of Technology Forecasting

The technology forecasting methods can be classified into two broad categories:

- Exploratory forecasting methods
- Normative forecasting methods

Exploratory forecasting considers a technological change on the basis of present level of knowledge and the pressures of competitive market. Historical growth of heights of buildings, speed in automobiles, substitute material for rubber, wood, etc., can be analyzed with an exploratory forecasting approach. Exploratory technological forecasting starts from existing basis of knowledge and aims at predicting future. Explorative forecasting is focused on predicting evolution of a new technology on a pre-determined curve. Normative forecasting considers technological change as a response to economic and societal requirements. In normative forecasting, technology development takes place as per the needs of the society. Normative forecasting in a way provides what is likely to be required in the future. It is based on the assumption that the future of technology is dependent on aspirations and changes in society's aspirations, expectations, needs and other factors like economic, political, etc. Normative technological forecasting methods work backwards. They first assess future goals, needs, desires, mission, etc., and from terminal point work backward to present. Normative forecasting is more pro-active. Table 4.2 lists a number of technology forecasting methods, which will be discussed subsequently.

Table 4.2 Technology forecasting methods

Technology Forecasting Methods	
Exploratory Methods	Normative Methods
Delphi Method	Relevance Tree
Extrapolation	Morphological Analysis
Growth Curves	Mission Flow Diagram

A) Delphi Method

Delphi method is one of the most widely used exploratory methods for technology forecasting. The Delphi method is a structured process for gathering knowledge from a group through a series of questionnaires with controlled opinion feedback to obtain forecast. Delphi method relies on a meaningful communication among a group of experts for formation of a group judgment. The Delphi method was originally developed at the Rand Corporation. It derives its name from the Oracle of Delphi, a prophet in ancient Greece. Development of the Delphi method started in 1944. General of US Air Force, Henry Harley "Hap" asked Douglas Aircraft company to establish in 1946 a

Project RAND (an acronym for research and development) to study the 'broad subject of inter-continental warfare other than surface'.

In 1959, Helmer and fellow Rand researcher published a paper title 'The Epistemology of the Inexact Sciences', which paved the way for development of Delphi method. The Delphi method recognizes human judgment as necessary and useful inputs for forecasting future. Assessment of the direction of long-range trends, with special emphasis on science and technology and their probable effects on society was one of the very first applications of the Delphi method at the RAND Corporation. The study covered six topics:

- Scientific breakthroughs
- Population control
- Automation
- Space progress
- War prevention
- Weapon systems

The first Delphi applications were aimed at forecasting new technologies, likely inventions and the social/economic impact of technological change. Over the period, application of Delphi method extended to industry, government and academics and also beyond technology forecasting. It is sometimes used for normal enquiry on issues related to public health, education, etc.

The objective of most Delphi applications is creative exploration of ideas and information on technology and related issues to help strategic decision making. Following are the objectives of the Delphi method:

1. To develop a range of possible alternatives
2. To explore or expose underlying information resulting in different judgments
3. To seek out information, this may generate a consensus on the part of the respondent group
4. To correlate informed judgments on a topic spanning a wide range of disciplines

The Delphi method considers that the experts in the given technology are the best sources to give opinion on technological forecasts. Delphi method works by relying on expert's opinion for technology forecasting. At the heart of Delphi is the group communication amongst geographically dispersed panel members. Thus, the Delphi method is an exercise in group communication among a panel of geographically dispersed experts. Instead of using services of one expert, however distinguished he may be, opinion of a committee of experts is sought. The Delphi approach involves participation of experts anonymously. Delphi method facilitates experts to arrive at a consensus to complex issues. This is achieved through a simple and structured Delphi methodology. The methodology involves the use of questionnaire to get opinions and views of panel expert on some issue or problem. Delphi method eliminates the disadvantages associated with conventionally used committee approach to find solutions.

There are three essential elements of the Delphi method:

- Panel of experts
- Anonymity of experts
- Structured retrieval of ideas and information with feedback process

Delphi method comprises of a series of questionnaires sent to a pre-selected group of experts. The questions are sent normally via email. These questionnaires are designed for the following purposes:

- To elicit and develop individual responses to the problems posed and
- To enable the experts to refine their views as the group's work progresses in accordance with the assigned task

Delphi method is characterized by three aspects: anonymity, controlled feedback and statistical response. The coordination of Delphi method is normally conducted by one individual. He is known as the director. A typical panel for Delphi method of forecasting consists of between ten and fifty members. The members of the panel are selected based on expertise and experience as well as availability and security considerations (e.g., commercial or military). In Delphi method, approach followed is iterative while eliciting expert's opinions. Each iteration is called a round. In each round, the members are asked for their views on the probability (likelihood) and timing of the occurrences of future events. Direct interactions among panel members are not allowed. Feedback to all panel members is provided. At the end of each round, the results (responses) are summarized statistically and circulated among panel members. In the first round, normally member's response varies widely. However, as members anonymously communicate their rationale, in successive rounds a consensus is reached. The director subsequently consolidates the results of this final round, which constitute the reported forecast.

B) Extrapolation

Extrapolation and trend analysis methods use historical data to gain insight into future technologies and development. The assumption in this method is that the future is an extension of the past and future is based on past. Hence, by identifying and extrapolating the appropriate trends from the available data, future can be predicted. In this type of forecasting, the factors that shaped the historical trends must be very carefully considered. In trend extrapolation, data is analyzed with known relevant trends. These trends can be extended in time to predict capability. Several approaches to trend extrapolation have been developed over the years. These methods examine trends and cycles in historical data. Then, a suitable mathematical/quantitative technique is used to extrapolate to the future. The basic assumption of these methods is that the factors responsible for creating the past events will continue to operate in the future.

For forecasting short term horizons, this assumption is often valid. However, this assumption does not seem to be normally working when creating medium and long term forecasts. In determining appropriateness of application of trend extrapolation model, stability of the environment is one of the key factors. Availability, reliability and suitability of historic data are of vital importance in these studies. The study of the historical data is called exploratory data analysis. The purpose of exploratory data analysis is to identify the trends and cycles in the data. Based on the trends, appropriate model can be chosen. There are many mathematical models for forecasting trends and cycles. Choosing an appropriate model for a particular forecasting application depends on the historical data. There are two types of extrapolation based on the rate of progress of the past behavior:

- Linear method and
- Exponential methods

Linear extrapolation: It is used where the relationship between the variables is linear, i.e., change in value of a variable P will result in change in value of variable Q to aQ, where 'a' is a constant. The trend can be explained with the help of the following linear equation:

$$Yi = a\,Xi + b$$

where

- Yi: value of the dependent variable in the ith time period
- Xi: value of the independent variable in the ith time period
- The values of a and b are estimated by the method of sum of squares and minimizing them from the projected extrapolation

Trend extrapolation: It is the exponential method. The trend is explained with the following equation:

$$Yi = a\,b^{Xi}$$

where

- Yi: value of the variable to be estimated
- XI: is the impact variable
- a,b: constants

There can be a situation when trend does not follow both linear method and exponential method. In such cases, other appropriate equation close to the trend can be used.

C) Growth Curve

A growth curve is an empirical model, derived from experimentation rather than theory, about the evolution of a quantity over time. Growth curve has early application in biology for growth of one variable such as weight, population, size, body height or biomass plotted against time.

Technologies go through following phases:

- Invention phase,
- Introduction and innovation phase,
- Diffusion and growth phase, and
- Maturity phase

During these phases, growth of the technologies is similar to the S shaped curve called S-curve. This growth curve forecasting method is used to decide the upper limit of performance for a specific technology.

For instance, Figure 4.2 shows the number of telephones per 1000 population as a function of time. The upper year of diffusion is determined as one phone per person. The year in which the upper limit is reached can be extrapolated from the S-curve, and it occurs between 1990 and 2000.

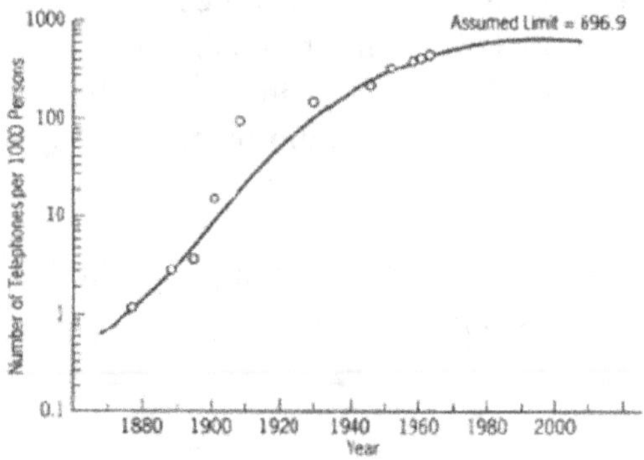

Figure 4.2 Growth curve example.

There are basically two types of S-curve formulations, which can be adopted based on the requirements of the forecaster:

- Pearl - Reed Curve
- Gompertz Curve

C.1) Pearl - Reed Curve

Pearl - Reed curve is named after United States demographer Raymond Pearl. He popularized its use for population forecasting. The pearl curve is used to track individual technologies. This curve is also known as logistics curve.

The formula for Pearl - Reed curve is as given in the equation below:

$$y = L/(1 + a * e^{-bt})$$

where

- Y: forecast variable
- L: upper limit to the growth of the variable y

- e: base of the natural logarithms
- t: time
- a, b: are the location and shape coefficients obtained by fitting the curve to the data

C.2) Gompertz Curve

These curves are used for the technologies where the initial stage growth is faster than in the Pearl curve. Gompertz curve is named after Benjamin Gompertz, an English actuary and mathematician. The Gompertz curve is not symmetric about the inflection point. Log of L/y does plot a straight line on a semi-log graph. The formula for Gompertz curve is given below:

$$Y = L * (e^{-bt})^{-kt}$$

where

- y is the variable representing performance
- L is the upper limit
- e is the base of the natural logarithms
- t is time, and
- b and k are the coefficients obtained by fitting the curve to the data

The choice of which curve to adopt depends on the dynamics of the technology to be modelled. Generally, the Pearl curve is used in the forecasting technology substitution. Gompertz curve is mostly applied to forecast absolute technical performance.

D) Relevance Tree

Relevance tree is a normative forecasting method. It starts with future needs or objective and subsequently seeks to identify the circumstances, actions, technologies, etc., required to meet them. A 'relevance tree' is a technique that subdivides a broad subject into increasingly smaller sub-subjects. Relevance tree is a pictorial representation with a hierarchical structure showing increasingly in detail finer levels. A relevance tree subdivides a broad topic into increasingly smaller subtopics. It identifies 'all' possible paths to the objective and provides a forecast of associated costs, durations and probabilities for each element. It can be used to identify problems and solutions, establish feasibility, select the 'optimum' solution and deduce the performance requirements of specific policies, technologies, etc. It may also be used to estimate the overall cost and duration of implementing policies or increase technological performance, thereby scheduling the detailed R&D program. A relevance tree maps all possible paths leading to an end result. In a decision tree, all possible paths are drawn resulting from a sequence of decisions. In a sense, relevance tree could be thought of as the reverse of a decision tree.

Relevance trees can be used to analyses situations with distinct levels of complexity. Relevance tree can be used for variety of purposes from forecasting a technology, selecting projects for a research program to designing a budget. It can be used to identify problems and solutions. It can establish feasibility and select the 'optimum' solution. It can suggest the performance requirements of specific technologies. It may also be used to estimate the overall cost and duration of implementing policies. One of its application can be scheduling the detailed R&D projects.

E) Morphological Analysis

Like relevance trees and mission-flow diagrams, morphological models are a normative technique. Morphological analysis is a method for exploring all the possible solutions to a complex problem. As a problem structuring and problem solving technique, morphological analysis was designed for multi-dimensional and non-quantifiable problems where causal modelling and simulation do not function well or at all. 'Morphological analysis' is a complementary technique, often used in conjunction with a relevance tree. The morphological model breaks a problem, such as technology road mapping, into parallel parts. Each part is treated independently. If more than one solution can be found for each part, then the total number of solutions to the entire problem is the product of the number of solutions to individual parts. The higher the number of potential solutions, the more promising is the technology.

The term morphology comes from classical Greek (morphe). Morphe means the study of form or shape. Morphological analysis deals with the structure and arrangement of parts of an object and how these parts fit together to create a whole. The 'objects' can be anything, e.g., physical, social or mental.

General morphological analysis was developed by Fritz Zwicky, the Swiss astrophysicist and aerospace scientist based at the California Institute of Technology (Caltech). Zwicky applied this method to such diverse fields like:

- Classification of astrophysical objects,
- Development of jet and rocket propulsion systems, and
- Legal aspects of space travel and colonization.

Morphological analysis was first applied by Fritz Zwicky to the aerospace industry. Zwicky chose to analyze the structure of jet engine technology. He first defined important parameters of jet engine technology. This included fuel type, oxidizer and thrust mechanism. He broke down each of these technologies into its component parts. Having exhausted the possibilities under each parameter heading, he assembled the alternative approaches in all possible permutations. For example, a ramjet that used atmospheric oxygen and a solid fuel. For some permutations, no systems or products were available. For other permutations, a jet engine system already existed. The permutations representing 'empty cells' are viewed as stimuli for creativity. For each such cell a question should be asked, "Why not?" For example, "Why not a nuclear powered ceramic fanjet?" Morphological analysis has been developed as a systematic procedure for collecting, counting, indexing and identifying all possible alternatives, to achieve some technological capability. Morphological analysis involves mapping a discipline to obtain a wide perspective of existing solutions and future possibilities. Zwicky outlined five basic steps in the development of the morphological method. The steps involved in morphological analysis are presented in Figure 4.3.

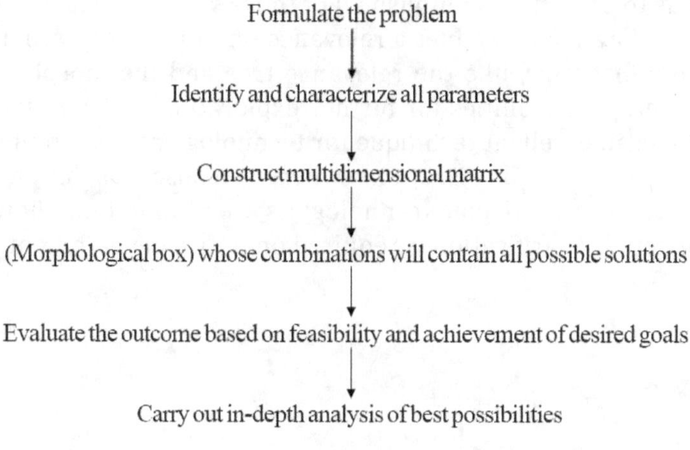

Figure 4.3 Steps in morphological analysis.

The primary use of morphological analysis has been in technological forecasting and new product ideation. Morphological models are very appropriate to a decision-making context, if one assumes that any one solution is as good as the others. The morphological analysis is normally much less precise in forecasting results. This is because it treats all possibilities as being equal, which is very unlikely. It does not pre-suppose a certain hierarchy in possibilities.

F) Mission Flow Diagram

Mission flow diagram is a normative method used for forecasting. Mission flow diagram was developed by Harold Linstoneas. It was originally used for analyzing military missions. Mission flow diagrams are a systematic way of identifying all possible future scenarios. Mission flow diagrams involve mapping all the alternative sequences by which a task or a function can be accomplished. The analyst has to identify significant steps on each route. Challenges at each route have to be determined and associated costs need to be estimated. The performance requirements can then be derived for each associated technology. In mission flow diagram, all possible contingencies are mapped. Thus, they are similar in many respects to decision trees. A decision tree charts out possible outcomes resulting from decisions taken in some sequential order. Similarly, a mission flow diagram identifies all paths and forks in the path, resulting from a sequence of events. Suppose the objective is to build a rail metropolitan system. Then, one has to lay out everything from today to the time operation begins. This process allows us to say what has to occur to accomplish all of those steps and stages. A mission flow diagram resembles a relevance tree discussed earlier. Mission flow diagrams, like the relevance tree and the morphological box, open up possibilities for further exploration. The mission flow diagram is an excellent technique for technology road-mapping. It is particularly useful in situations where a consensus has already emerged that a particular technology is worth pursuing, however, certain details/clarification is required on future development of the technology.

5

Technology Assessment

5.1 Introduction

Our world and our society are profoundly influenced by technology. It has helped raise our standard of living, but has also brought about particular problems. As prosperity increases, technology penetrates into more and more diverse areas. Accessible 'mechanical' technologies are increasingly being replaced by 'inaccessible' electronic and chemical technologies. Apart from ever present fear that technology could cause unemployment, there is also a more recent fear that it could destroy the fundamentals of our lives. Fear and uncertainty about the consequences of new technologies are not only social problems, they also present a considerable challenge for policy makers who can at least influence developments through legislation and administration (research, technology and industrial policies). The complex consequences of new technologies are not readily understood by the layman and experts are also less unanimous in their opinions. This is inevitable given the prognostic nature of technology assessment and also the different yardsticks and approaches: those who are looking for the benefits of new technologies (e.g. new drugs) will have a very different attitude to gene technology experiments compared with those who are fearful of the consequences of releasing genetically modified organisms or of manipulation of the human genotype.

Technology assessment attempts to provide policy makers with a rational basis for their decisions. It does not provide 'super expert opinions', but points out areas where specialists are in general agreement, where controversy exists, what assumptions or fears lie behind the differences of opinion and what risks are associated with the various possible options. It highlights facts that policy makers can assume to be accepted and the problems that they have to solve through political decisions or value judgments. By focusing on political discussion of relevant questions, the decision making process can be made simpler, more objective and efficient. Technology assessment provides input for the elaboration of technology policies which are not oriented exclusively towards technical efficiency and economic rationality but which also consider the social and ecological consequences of the introduction of specific technologies. The classic aim of technology assessment is to identify technology-induced risks early enough, to analyze in detail the range of possible social, economic, legal, political, cultural and ecological effects, to process investigation results in a problem-oriented manner, to present alternative decision-oriented options and at the same time to point out the various social interests and value judgments linked with the development and use of new technologies. New insights into the nature of technological change have stimulated the further development of technology assessment (TA) concept. It is based on the realization that technical developments are ultimately influenced by society and not determined solely by their own logic. This calls for a wider vision of TA, one which takes account of design-related factors and forces in the research and development (R&D) process and their application in a specific technology and which also allows social discussion of the objects and options of technological development. Thus, the core analytical functions of technology assessment are supplemented by constructive design and discussion-provoking functions.

They are reflected in the terms 'constructive', 'discursive' or 'participatory' technology assessment. Due to the rapidly changing nature of technology, new and complex ethical issues are emerging which bring into question the ability of society to address and hopefully resolve them. These new issues are arising in such areas as biotechnology, information technology, nanotechnology and nuclear technology and range from protecting the health and welfare of the public and the environment, to patenting living organisms and labeling products containing genetically modified organisms, to concerns regarding the alteration of the ecology of life. The problem is to a large extent and intensity of the conclusions of a series of questions. This is because the technology is a broad perspective, including a large number of areas, because of environmental conditions of the problems of millions of people around the world.

5.2 Technological Impact on Society

Technology and society or technology and culture refer to cyclical co-dependence, co-influence, co-production of technology and society (technology upon culture and vice-versa). This synergistic relationship occurred from the dawn of humankind, with the invention of simple tools and continues into modern technologies such as the printing press and computers. The academic discipline studying the impacts of science, technology and society and vice versa is called science and technology studies.

A) Values: The implementation of technology influences the values of a society by changing expectations and realities. The implementation of technology is also influenced by these values. In general. there are three major and interrelated values that influence and are influenced by technological innovations.

B) Mechanistic World: Viewing the universe as a collection of parts (like a machine) that can be individually analyzed and understood. This is a form of reductionism that is rare nowadays. However, the 'neo-mechanistic world view' holds that nothing in the universe is such which cannot be understood by the human intellect. Also, while all things are greater than the sum of their parts (e.g., even if we consider nothing more than the information involved in their combination), in principle, even this excess must eventually be understood by human intelligence. That is, no divine or vital principle or essence is involved.

C) Efficiency: A value, originally applied only to machines, but now applied to all aspects of society, so that each element is expected to attain a higher percentage of its maximal possible performance, output or ability.

D) Social Progress: The belief that there is such a thing as social progress and that, in the main, is beneficent. Before the industrial revolution and the subsequent explosion of technology, almost all societies believed in a cyclical theory of social movement and indeed, of all history and the universe. This was, obviously, based on the cyclicity of the seasons and an agricultural economy's and society's strong ties to this cyclicity. Since much of the world is closer to their agricultural roots, they are still much more amenable to cyclicity than progress in history.

E) Lifestyle: In many ways, technology simplifies life. It also includes/means:

- The rise of a leisure class

- A more informed society
- Sets the stage for more complex learning tasks
- Increases multi-tasking (although this may not be simplifying)
- Global networking
- Creates denser social circles
- Cheaper prices
- Greater specialization in jobs

F) International: Technology enables greater knowledge of international issues, values and cultures. Mostly due to mass transportation and mass media, the world seems to be a much smaller place. Other factors include:

- Globalization of ideas
- Embedding of values
- Population growth and control

G) Environment: Technology provides an understanding and an appreciation for the world around us. Most modern technological processes produce unwanted byproducts in addition to the desired products, which are known as industrial waste and pollution. Though most material waste is re-used in the industrial process, however, many forms are released into the environment, with negative environmental side effects, such as pollution and lack of sustainability. Different social and political systems establish different balances between the value they place on additional goods versus the disvalues of waste products and pollution. Some technologies are designed specifically with the environment in mind, but most are designed first for economic or ergonomic effects. Historically, the value of a clean environment and more efficient productive processes has been the result of an increase in the wealth of society, because once people are able to provide for their basic needs, they are able to focus on less-tangible goods such as clean air and water.

5.3 Technological Impact on Business

Changes in the technological environment have had some of the most dramatic effects on business. A company may be thoroughly committed to a particular type of technology and may have made major investments in equipment and training only to see a new, more innovative and cost-effective technology emerge. Indeed, the managing director of a multinational organization manufacturing heavy machinery once said that the hardest part of his job had nothing to do with unions, pay or products, but with whether or not to spend money on the latest technologically improved equipment. Computer technology has an enormous impact on education and health care. The advancements in medical technology, for example, have contributed to longevity in many societies. In addition, the introduction of robots in many factories has reduced the need for labor and the use of VCRs and microcomputers has become a common place in many homes and businesses. Unfortunately, there is a negative side to technological progress too. The introduction of nuclear weapons, for example, has made the destruction of the human race a frightening possibility. In addition, factories using modern technologies have polluted both air and water and contributed to various environmental and health-related problems.

Technology is a critical factor in economic development. Due to the advances of international communication, the increasing economic interdependence of nations and the serious scarcity of vital natural resources, the transfer of technology has become an important preoccupation of both industrialized and developing countries. For many industrialized countries, the changes in the technological environment over the last 30 years have been immense, particularly in such areas as chemicals, drugs and electronics. It is vital that organizations stay abreast of these changes not only because this will allow them to incorporate new and innovative designs into their products, but also because it will give them a firmer base from which to anticipate and counteract competition from other organizations.

High technology has become like a force of nature. It transforms the economy, schools, consumer habits and the very character of modern life. Investors pour money into it; parents urge their children to study it; communities vie to attract its factories; decorators adopt it as a style; politicians push it as a panacea, etc. For example, when the Gillette company developed a superior stainless steel razor blade, it deduced that such a superior product might mean fewer replacements and sales. Thus, the company decided not to market it. Instead, Gillette sold the technology to Wilkinson, a British garden tool manufacturer, thinking that Wilkinson would use the technology only in the production of garden tools. However, Wilkinson Sword blades were subsequently introduced and sold quickly. The transfer of technology is essential for attaining a high level of industrial capability and competitiveness. Multinational corporations are playing an increasingly important role in technology transfer because they invest abroad to expand production, marketing and research activities. There is also a growing consciousness amongst governments of the need to increase technology transfer to the developing countries to help stabilize their economic and social conditions.

In spite of the many differences in social, political, cultural, geographic and economic conditions, there are some common characteristics in the technological environments of developing countries. The most common technology transfer from industrialized to developing countries has been in agriculture and health care. As a result of improved health care systems, infant mortality rates have been cut while the incidence of once common diseases such as malaria and typhoid has been reduced in Latin America, South-East Asia and Africa (although the incidents of the AIDS virus has increased alarmingly). Similarly, agricultural technology has increased agricultural productivity in Brazil, India and elsewhere. However, in most developing countries, technology has made little impact on the productive systems, income distribution and living conditions of majority of the population.

Technology transfer is a complex, time-consuming and costly process and the successful implementation of such a process demands continuous communication and co-operation between the parties involved. Furthermore, technology transfer cannot be effective if it experiences conflict with the economic and social needs of the recipient country. The agricultural development of north-eastern Brazil, for example, was largely financed by international banks and financial organizations in the 1960s. Much of this region had been inhabited by Brazilian aborigines, but it was owned by a small number of wealthy landowners. However, the introduction of large-scale mechanical agricultural technology in areas of the tropical rain forest of the Amazon has caused serious environmental damage such as erosion of tropical top soil and the destruction of the natural environment of numerous birds and animals and has displaced a large number of the local inhabitants of the forests. Technology transfer may also become a serious source of conflict between donor and recipient countries. The recipient country may feel that the donor is trying to dominate it through technology, capital and production. Dependence on foreign technology can be viewed as a serious threat to economic independence. Countries that export technology may experience different problems. For the seller of technology, the technology transfer can result in unemployment in the home country and future loss of technological superiority. For example, Japan transferred modern steel production technology to South Korea in the early 1970s. As labor and production costs in Japan increased, the Korean steel industry began to take over a significant portion of the previously Japanese controlled international market. Some Japanese executives are now complaining that the cost of technology transfer has been much greater than the income received through the sale of technology.

Technology can be transferred from person to person, industry to industry and government to government, although the government of any country generally plays the most important role in facilitating or impeding the transfer process. Contacts amongst students from different countries are also a means of technology transfer as are journals, books, technical and professional publications, trade magazines and product pamphlets. Furthermore, multinational corporations play an important role in technology transfer by transferring information and technology from the parent company to subsidiaries in other countries, training foreign employees, etc.

The future development of the international business environment depends on numerous factors, including the political and economic environment, the development of technology, population growth, energy availability and natural resource depletion. It can be expected that the gap between the rich and the poor countries will increase and that living conditions in many poor countries will deteriorate even more. The rich countries will be obliged to extend technology to the poor countries. Technology has had a tremendous impact upon the global business environment. Communication, transportation and production efficiency are various areas of business which have been enhanced by the development and improvement of technology. As continual enhancements are made, the world continues to 'grow smaller' and businesses have further reach than ever.

A) Computers: The most important technological development to impact the global business environment is the world of computers. There are various programs which help maintain records of inventories and shipments. Email allows for instantaneous communication almost anywhere in the world. Besides its speed, email is easily forwarded and retained. The communication in the global business environment is improved with the use of email. The impact of computers on the global business environment is wide-ranging and also includes internet, which is a useful tool for international companies. By using internet, companies across the world can perform research and learn more about partners and suppliers.

B) Conference Calls and Video Conferencing: Conference calls allow people in multiple locations to be involved in the same conversation. Video conferencing provides the same service, but with the added benefit of all parties being able to actually see each other. Both of these forms of communication have a definite impact on the global business environment. With either form of technology, a parent company in USA can have a conversation with a raw material supplier in Germany and a manufacturing plant either in India or in China. This improves communication on a global scale and enables all parties to understand specific plans and agreements.

C) Transportation: The shipment of raw materials and finished products is absolutely vital to any business, but particularly those with an international scope. Transportation technology enables a company on one continent to send its raw materials or products to another company in a different continent. Technological advancements in airplanes, cargo ships and railways allow for quicker and cheaper delivery, which impacts business by making global distribution more feasible.

D) Manufacturing Technology: Increased efficiency of manufacturing plants has a certain impact on the global business environment. By having the capacity to produce materials and products more quickly and efficiently, a company is able to produce quantities needed to supply global demand. Robotic technologies and factory lines have enhanced the speed at which materials and products are manufactured. For a company to be a player in the global business field, it must be able to keep up with demand.

E) Shipment Tracking: Corporations now have the ability to track shipments virtually anywhere across the world. Global positioning systems (GPS) allow accurate tracking. The implication of this technology on the global business environment is the ability to let customers know exactly where their shipments are at any given time. This technology creates secure relationships within the global business field. The impact of technology has been most evident on business within the last 20 years with the growth of the microchip and the adoption of personal computers, as well as the retooling of old systems and the development of new systems that rely on computer technology. Globally, we are in the age of technology, a time that history will record as a period when businesses made giant strides in creating smart machines, smarter processes and perhaps even smarter business people.

F) Technology in the Office: Mass adoption of the word processor and, later, the personal computer brought office productivity out of the dark ages. Gone are the days of using carbon paper and mimeograph machines to make a few duplicates. High-speed printers are capable of producing hundreds of pages of collated and stapled documents in 256 colors.

G) Technology Out of the Office: Cell phones beam from satellites positioned in a geostationary orbit 22,300 miles above our heads, allowing sales professionals on the road to communicate with customers, office or home. Wireless internet lets businesspeople access their mailbox from a cellphone or manage their calendars. PDAs (personal data assistants) now keep contact numbers and appointment details and dates, as well as provide the latest updates on the stock market.

H) Technology in Manufacturing: The impact of technology on business is especially evident in the manufacturing sector. There are few mass-produced products that are manufactured without the aid of computer automation. Computers integrated with processing machines in manufacturing companies have greatly increased the number of widgets a company can produce. Computer sensors can scan finished products and detect defects that might be missed by the human eye.

I) Technology Expands Media: Media outlets for business marketing have increased as both broadband and wireless internet adoption has exploded among all consumer groups. The fear that multicultural groups were being left behind in a so-called digital divide was only recently refuted when an April 2009 study by Pew Internet and American Life revealed that minorities were using the wireless devices like mobile phones to access the internet rather than access it using broadband.

J) Technology in Biotech Sector: Technology in the science sector has revolutionized biotech and pharmaceutical industries as they peer deeper into the cause of diseases like cancer and HIV. Patents for new drugs and additional indications for existing drugs have been a boon to industry and have positively affected people's health and life expectancy. Information, communication and computer technology have already changed the business world in several ways. Laptops and mobile devices offer more constant and efficient methods of communication that aid businesses in improving their networking skills in a global market. New software and advanced applications help companies improve the way they conduct their research and develop their products and services. Wi-fi connectivity is the latest means of communication between 'n' numbers of individuals.

K) Accessibility: Communication and mobile technology has introduced mobile devices like tablet computers and personal digital assistants (PDAs) like palmtops to the workforce, enabling employees to conference and chat with peers and customers anywhere in the world instantly. With a drastic increase in both the frequency and speed in which businesses can communicate and receive feedback from customers, contact peers and learn more about their competitors around the world, companies have become more accessible than ever.

L) Application: Many businesses have maintained websites for years and more continue to join, aware that this technology increases global awareness of their existence. Now, more companies are also utilizing recent advances in social networking technology by creating profiles on websites like Facebook, LinkedIn and Twitter. These tools help them share industry-related news and stay in direct and constant contact with their clients. Thanks to the interactive technology these websites provide, the businesses receive immediate feedback from customers, thus, increasing the speed of their research and development.

M) Expert Insight: According to a study by the Center for Information Systems Research at the Massachusetts Institute of Technology, the three specific perspectives in which technology can be viewed in regard to the business world are those of improving a business, surpassing the competitors and deciding whether or not to enter a specific business industry.

5.4 Understanding Technology Assessment

A) Meaning of Technology Assessment: It means estimating the effect of new products and processes on the operations of a firm, other firms, society and the environment. It also includes review of product vision, features and users and compare with the operating systems, development tools and software techniques that are available today and in the future. It also encompasses matching the software design to the vision. TA is the study and evaluation of new technologies. It is based on the conviction that new developments within and discoveries by the scientific community are relevant for the world at large rather than just for the scientific experts themselves and the technological progress can never be free of ethical implications. Technology assessment recognizes the fact that scientists normally are not trained ethicists themselves and accordingly ought to be very careful when passing ethical judgment on their own or their colleagues, new findings, projects or work in progress. Technology assessment assumes a global perspective and is future-oriented rather than backward-looking or anti-technological. Scientific research and science-based technological innovation is an indispensable prerequisite of modern life and civilization. There is no alternative to it. An important problem which TA has to deal with is the so-called Collinridge dilemma, where on one hand, impacts of new technologies cannot be easily predicted until the technology is extensively developed and widely used, whereas, on the other hand, control or change of a technology is difficult as soon as it is widely used.

B) Purpose of Technology Assessment: Usually the technology assessment or measurement of its impact covers safety, cost, efficacy, effectiveness and efficiency (cost-benefit, cost-effectiveness or cost-utility) of a technology and the social, legal and ethical consequences of its introduction. This information provides a basis for rational decision-making whether or not to exploit a given technology. For example, the purpose of technology assessment in healthcare is to produce information on the impact of the introduction and use of a specific technology on citizens, patients and on the health care organizations including their funding. The objectives of technology assessment are as follows:

- To assess the strengths of a university or research institute or typically, a department within a university or research institute. This may include technologies, equipment and most importantly the expertise of researchers.
- To identify opportunities to promote the expertise of the department being assessed. These new opportunities may lead to increased direct and indirect income for the department.
- To be a source of strategic management information for the department.
- To motivate staff and increase awareness of the department.
- To support local technology transfer (inwards and outwards) and technology commercialization, particularly involving small and medium sized companies.

Technology assessments may also be used by companies to identify important resources that are often overlooked, e.g. auditing of internet protocol (IP).

Information gathered in technology assessments may include know-how, documents, processes, techniques, skills, equipment and computer software. Apart from serving as a useful management tool, the information can improve internal communications, prevent duplication of effort by promoting internal technology transfer, increase employee awareness of new opportunities and enable employees to feel recognized and rewarded for their technical contributions. Conducting an audit will raise expectations that something positive is going to happen as a result. If these expectations cannot be fulfilled, the audit may have a negative effect.

5.5 Process of Technology Assessment

Technology assessment involves iterative cycles of planning, gathering data and reporting. It is best to plan the assessment at the same time that the technology introduction is being coordinated, so that necessary steps can be taken such as collection of baseline data. All program assessments follow planning, gathering data and reporting results, the process of which consists of the following steps, as also outlined in Figure 5.1.

A) Planning:

A.1 Describe the technology & context:

- Use in terms of its function or purpose, expected effects and resources needed.
- Describe the context such as characteristics and objectives in relation to the technology.

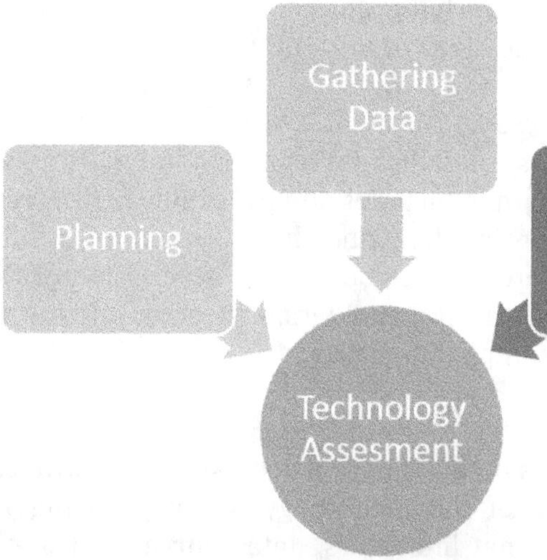

Figure 5.1 Technology assessment process.

A.2 Identify stakeholders and their needs

Stakeholders are the individuals and organizations involved in developing and supporting the technology, those served or affected by the technology and the intended users of the assessment. The needs of stakeholders generally reflect the central questions the stakeholders have about the technology. Determining stakeholder needs helps to focus the assessment process so that the results are of the greatest utility.

A.3 Determine the assessment purpose

Identifying a clear purpose helps determine how to conduct the assessment. Three general purposes for instructional assessments are to gain insight, change practices and/or measure effects.

- Gain insight - to clarify how the activities should be designed to bring about expected changes.

- Change practice - to improve the quality, effectiveness or efficiency of the technology.
- Measure effects – to examine the relationship between technology, application and effect.

A.4 Identification of intended use
Intended uses are the specific ways assessment results will be applied. They are the underlying goals of the assessment and are linked to the central questions of the study.

A.5 Creation of an assessment plan
The assessment plan is a detailed description of how the assessment will be implemented that includes identification of the resources available for implementing the plan, what information is to be gathered, the research method(s) to be used, a description of the roles and responsibilities of the stakeholders, instructor and/or assessor and a timeline for accomplishing tasks.

B) Gathering Data:
Data gathering focuses on information acquisition that will convey a well-rounded picture of the technology. Data gathering includes consideration about what indicators, data sources and methods to use, the quality and quantity of the information, human subject protections and the context in which the data gathering occurs.

C) Reporting Results:
C.1 Analyze data
Data analysis involves identifying patterns in the data, either by isolating important findings (analysis) or by combining sources of information to reach a larger understanding (synthesis) and making decisions about how to organize, classify, interrelate, compare and display information. These decisions are guided by the questions being asked, the types of data available and input from the stakeholders.

C.2 Conclusions and recommendations
Conclusions are linked to the evidence gathered and judged against agreed-upon standards set by stakeholders. Recommendations are actions for consideration that are based on conclusions, but go beyond simple judgments about efficacy or interpretation of the evidence gathered.

C.3 Reports

Factors to consider when reporting results, or dissemination, include tailoring a report content for the audience, explaining the focus of the study and its limitations and listing both the strengths as well as weaknesses of the technology.

Technology assessment uses an interdisciplinary approach in its systematic and comprehensive analysis of the possible consequences of the use of particular technologies (not necessarily 'new' ones). On the basis of the information gained in this way, options can be devised for those involved in policy-making and the control of technology. Particular attention is paid to the analysis of unintended and generally long-term secondary or tertiary effects. Due to the nature of the object of research and the interdisciplinary composition of the research team, a variety of quantitative and qualitative methods are used: brainstorming, literature research, document analysis, expert consultation, case studies, cross-impact analysis, cost/benefit analysis, computer simulations, scenario development and also certain processes calling for public involvement. The methods must be adapted to the subject under investigation.

There is no such thing as 'technology assessment method', because it is not a single method, but the variety of methods or method mix and specific procedure that characterize a successful investigation, as:

- Define problem
- Details of the technology
- Mapping/judgment about technological developments in future
- Evaluation of impact on social and individuals
- Future social developments
- Identify, analyze and evaluate consequences
- Account for political options available
- Communicate the results

5.6 Technology Audit

A technology audit is a method of investigation aiming at the evaluation of the (a) technological capacity, (b) procedures and (c) needs of an SME or an organization. Furthermore, it is a method of identifying the strong and weak points through the characterization and general assessment of the firm's basic know how (marketing, management, finance, human resources, etc.). It is a process of analysis, which leads to concrete proposals – is also known as 'action plan'. From information technology point of view, technology audit is concerned with the information technology, how it can be controlled and examined as the entire information technology is used for the purpose of infrastructure. This is mainly being used by the people who work in organizations and have needs to carry out different technological audits. It has been proved to be quite helpful for the information technology departments as it gives these departments an opportunity to gather a wider range of information. Moreover, it has been found that automated data processing is another name of technology audit. Technology audit is also being named by some as information technology audit or the information systems audit. Using it in either way generates the same result. These have also been termed as computer audits. These have been termed so in order to find out how the things within the company are helping and being utilized in the audit department, which basically is the checking of the accounts that had been maintained till now. The technological audit can only be carried out by the information technology experts. They are the ones who can easily respond to such activities. Furthermore, it has also been developed for the purpose of finding out whether the assets are being utilized in the proper way and all of it needs to be catered and safeguarded with the help of information technology. Another thing which must also be taken into account is whether the users had been able to achieve their entire objectives or not. The technology audit at times can also be performed in the form of financial statement or other audit which helps in determining the performance of the company.

The technology audit is carried out by external consultants in close collaboration with the management and personnel of the company. It is based on the structure:

Data Collection ➡ Analysis ➡ Synthesis ➡ Report

After successful completion, it will result in an action plan leading to technological improvement, acquisition of needed technologies and/or expertise, etc. It is a good starting point for further development of business provided it is carried out by experienced consultants and taken seriously by the management of the company.

It should be stressed that the technology audit itself cannot solve fundamental problems, provide immediate benefit, rectify faults and be a substitute for organizational difficulties. The general aim of the technology audit is to evaluate the capacity of firms and organizations to integrate new technologies, work with technological partners and better define what they need to successfully integrate these technologies into the company.

Technology audit is a need of organization towards innovation. Technology assessment involves iterative cycles of planning, gathering data and reporting. Thus, overall, we have seen in this unit that the concept of technology audit is a method of investigation aiming at the evaluation of the capacity, procedures and needs of an organization. It though itself cannot solve any problems, or provide immediate benefit.

6

Technology Diffusion and Absorption

6.1 Introduction

It is interesting to understand how the users react to various technological developments and how they actually use an array of products or services based on these technologies in their day to day life. Here users can be either an individual/group of individuals or organization. The process by which any new idea or technology is chosen or accepted by an individual or organization for use can be termed as technology adoption.

The technology adoption will then lead way to technology diffusion which is discussed in the subsequent topic. The process of acceptance of new idea can be divided into the following stages:

- **Awareness:** The user or industry comes to know about new idea or technology but does not have detailed knowledge about the same.
- **Interest:** At this stage, the user gets attracted towards the idea and tries to seek more information necessary to explore its usefulness.
- **Evaluation:** At this stage, the user mentally assesses the usefulness of the idea for prevailing situation and comes to a decision whether or not to actually use it.
- **Trial:** At this stage, the user actually tries the idea on experimental basis. This will provide inputs necessary to customize the idea to fit in the existing set up or system.
- **Adoption:** At this stage, the user is convinced about the usefulness of the idea and decides to continue the use of the same in future.

The information communication plays a very decisive role in acceptance of new idea or technology. There are multiple sources of information which can be communicated by internet, television, newspapers, magazines, peers or friends, various organizations or associations or visit to the place where new idea or product is actually working. The adoption of particular technology in any industry or organization is dictated by many factors like customer's expectation, requirement of capital investments, time required for acquisition of necessary skills, existing structure if any, prevalent or probable competition, regulatory framework, incentives, anticipated cost reduction or improvement in efficiency, potential of increase in market share, etc., to list a few.

6.2 Life Cycle of Technology Adoption

Lot of research has been conducted across the globe on these aspects of technology adoption. In 1957, Everett M. Rogers along with Joe M. Bohlen, and George M. Beal had developed technology adoption life cycle at University of Iowa while following the purchase of hybrid corn seeds by farmers. The model of technology adoption life cycle explains the adoption or acceptance of a new product or innovation. It is affected by demographic and psychological characteristics of defined adopter groups. Figure 6.1 illustrates the technology adoption lifecycle.

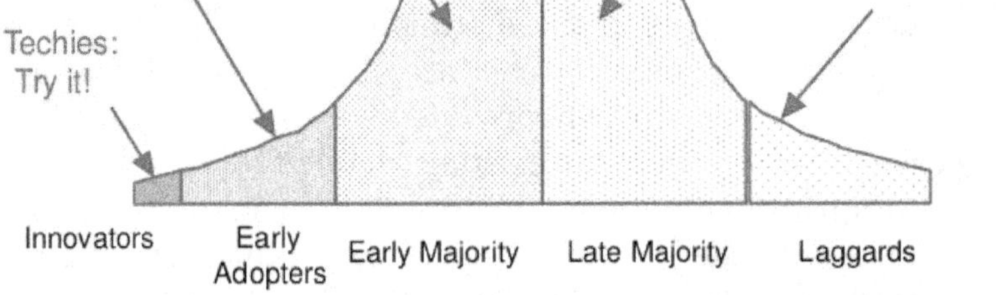

Figure 6.1 Technology adoption lifecycle.

A) Innovators: A person or an organization willing to try new ideas anticipating it will help to improve their performance can be termed as an innovator. With reference to the original research carried out at Iowa State University, innovators are active in the community, resource rich and willing to take calculated risk. In an early stage of technological developments when there is no guarantee of success or yielding favorable results after deployment of technology, innovators adopt that technology. As observed in any community, there is very little number of innovators.

B) Early Adopters: Early adopters can be a person or organization which is quick enough to follow the innovators. People in the early adopter category seem to have the greatest degree of opinion leadership in most social systems. They provide advice and information sought by other adopters about an innovation. The early adopter is usually respected by his or her peers and has a reputation for successful and discrete use of new ideas.

C) Early Majority: This is the next category of adopters and as seen in the figure, number of adopters start surging with joining of more and more persons or organizations. This group of adopters act as informal leaders and have limited resources than the innovators and early adopters.

D) The Late Majority: The late majority is a representation of skeptical adopters in a community and they adopt new ideas after the average member of a social system. The adoption in this case of adopters most of the times is not by choice but due to economic necessity and increasing social pressure. This group is vigilant about new ideas or products and is hesitant to adopt until most others in their social system do so first.

E) Laggards: The last group of adopters is termed as laggards. These adopters are adamant and accept the new idea or product or technology very last. They have negligible opinion leadership and are often isolated from rest of the community. They are fond of the customs and traditions and have very little willingness to break the customs in light of fresh knowledge. With the changing pace of technological developments, it is quite possible that the product or technology adopted by laggards may very well be on the brink of obsolescence.

6.3 Understanding Technology Diffusion

Various technology breakthroughs over the centuries played an important role in the industrial development and economic growth across the globe till date. The technology progress is an essential element for economic growth and it has already been proved by various research studies. However, technological breakthroughs or innovations do not immediately achieve economic significance. Time is required on the part of people to understand how these changes in technology or innovation will improve their efficiency and optimize the use of resources. This process in which an innovation is communicated through certain channels over time among the members of a social system can be termed as technology diffusion. It is a slow and continuous process. Technology diffusion is equally important as technology innovation. Unless the technology spreads across the users, it will not achieve the economic significance and impact the productivity. The process of technology diffusion is spread over a period of time and affected by many factors. As Hall & Khan (2002) pointed out, diffusion can be considered as the cumulative or aggregate result of a series of individual calculations that weigh the incremental benefits of adopting a new technology against the costs of change, often in an environment characterized by uncertainty (as to the future evolution of the technology and its benefits) and by limited information (about both the benefits and costs and even about the very existence of the technology). In the era of knowledge based economy, the technology dissemination can happen via various ways and with the necessary changes in the original technology time to time and depending on the type of industry.

The relative speed at which the technology spreads among various users (both individual as well as organizations) can be termed as rate of technology diffusion. During his study of the history of diffusion of innovations, Nathan Rosenberg (1972) revealed the two important characteristics of diffusion process as overall slowness and extensive deviation in the rate of acceptance of various innovations. It is important to understand why the diffusion process occurs at slow rate. This slowness of diffusion can be understood with following reasons as expressed in a study by Mulder *et al.* (2003):

- The uncertainty with respect to future of the technology and some degree of irreversibility in investment.
- The possibility of (expected) rival innovation or imitation of innovation by competitors.
- The fact that over time the performance of existing technologies improves and their price reduces due to learning and spill-over effects.

- A final reason highlights the role of vested interests. As switching to new technologies (temporarily) reduces expertise and destroys rents associated with working with relatively old technologies for particular subgroups in the economy, these groups may engage in efforts aimed at keeping the old technologies in place.

Shapira and Rosenfeld (1996) suggested following measures which can have a significant role in fueling the rate of technology diffusion in organizations especially in small and medium scale organizations:

A) Awareness-building and technology demonstration: This deals with spreading awareness amongst potential users by proving them more information about available technologies, their possible applications, their benefits and costs. For example, in India, various organizations or agencies like Indian Council of Agricultural Research (ICAR), Centre for Sustainable Technologies (CST) and Department of Science & Technology carry out various initiatives to create awareness regarding various technological developments in the related field.

B) Information search and referral services: These measures intend to decrease the information search costs associated with technology diffusion. Information services often add further value by qualifying information requests and matching user needs with appropriate resources. For example, National Small Industries Corporation Ltd. (NSIC) operates through countrywide network of offices and technical centers in India and offers information and other technical services to firms. In several countries, new initiatives are underway to use the internet as a medium to service technical information needs.

C) Technical assistance and consultancy: This encompasses a wide band of measures which support experts to assess business problems, identify opportunities to upgrade technologies and industrial practices and assist in implementation. These measures seek to address limitations of expertise among both users and suppliers of technology and to stimulate and assist firms to take action (or, in some cases, not to act on an undesirable investment). Technical assistance services are located in many applied technology centers.

For example, Micro, Small & Medium Scale Enterprises (MSME) Development Institutes are spread across all states of India and offer facilities for testing, preparation of project and product profiles, technical and managerial consultancy, assistance for exports, pollution and energy audits, etc.

D) Training for technology: The effective deployment of technology and improved operational techniques invariably involves changes in human capital requirements. A very common technology diffusion measure is, thus, training, conducted in many different forms, including on-the job training, classroom training, management seminars, team-building workshops and distance learning. These measures address the tendency of technology users to under-invest in human capital development, which often not only hinders the initial decision to deploy a technology but can also lead to subsequent inefficiencies once in use. Additionally, special measures to promote training for technology diffusion may also address deficiencies among existing institutions and vendors (who may be unable to effectively mount courses in new technologies without additional support). For example, Composite Applications Laboratory (CAL) setup under the aegis of the Department of Chemical Engineering, Indian Institute of Technology (IIT) Kharagpur conducts Summer/Winter School for catering to training needs of the composite industry.

E) Collaborative research and technology projects: To address the gaps between technology development and deployment, a range of collaborative public-private research mechanisms have been established. These measures also seek to shorten the time taken to commercialize new technological innovations and, through industry involvement, focus research on key needs and opportunities. Often, collaborative research efforts are embodied in the numerous applied technology centers now found throughout the Organization for Economic Co-operation and Development (and other) member countries. For instance, in India, Centers of Excellence (CoEs) aim at making businesses globally competitive by leveraging the public-private partnership model for development of cutting edge technologies. It includes project 'Next Generation Network & Network Technology' by Indian Institute of Technology (IIT) Kharagpur sponsored by Vodafone Essar and project 'Telecom Technology & Management' by IIT Delhi sponsored by Bharti Airtel. The Centre of

Excellence in Wireless Technology (CEWiT) has been set up under a public-private initiative with the mission of generating leaders in research, development and deployment of wireless technology. Similar centers and projects can also be noticed in other countries.

F) Personnel exchange and the support of R&D personnel: Potential users of technology, especially small or mid-sized enterprises may lack the internal expertise to absorb new technologies or they may lack resources to apply their existing personnel to new research and technology projects. In Japan, local public technology centers accept staff from smaller firms for periods of time to receive training in new technologies and participate in cooperative research. Increasingly, international exchanges are encouraged. In Germany, ministries have sponsored programs to subside research personnel in small and mid-size firms to help them absorb and develop new technologies.

G) Standardization: Standardization means a framework of agreements to which all relevant parties in an industry or organization must adhere to ensure that all processes associated with the creation of a good or performance of a service are performed within set guidelines. Uncertainty about the compatibility of a technology can present barriers to diffusion investments by users. The diffusion of technology can be accelerated by common agreement between technology developers and users about standards and technological compatibility. For example, in India for standardization of manpower or talent pool, NASSCOM has been working with the Indian ITeS-BPO industry players to create a national assessment and certification program, the NASSCOM Assessment of Competence (NAC), which becomes an industry standard. It ensures the transformation of a 'trainable' workforce into an 'employable' workforce. In a different context, the development of standard measures to document quality, through ISO 9000 and subsequent reference marks, has also facilitated the diffusion of quality measurement techniques and the avoidance of duplicative marks.

H) Financial support: These measures are intended to reduce financial constraints among users associated with the initial or ongoing costs of adopting new technologies. Measures can include direct financial support or subsidies to enterprises, through grants, loans, or interest write-downs. Other mechanisms are loan guaranties (often associated with third-party lending institutions), equity or near-equity investments and various kinds of royalty agreements. In many cases, public financial policies to promote technology diffusion operate through intermediary institutions, including banks and quasi-governmental corporations. Requirements may be set as to the user's own cost-share or match. For instance, technology development boards in many countries provide financial assistance to the industrial concerns and other agencies attempting development and commercial application of indigenous technology or adapting imported technology for wider domestic application.

I) Procurement: Purchasing and specification policies by public institutions and large private firms can have a role in promoting (or constraining) the diffusion of technologies. In some cases, the public support of large firm investments is associated with conditions for local procurement, which then may require supplier upgrading programs. This may also be in the form of joint-production or offset agreements typically found in military, aerospace or large transportation projects. Procurement policies may generate concerns related to free trade. A series of new programs have sponsored different forms of inter-firm collaboration to promote technology diffusion. The efforts seek to resolve common problems and share information and learning, achieve scale economies in service provision and technology deployment and strengthen ongoing business and technology development relationships. Collaborative efforts may be 'horizontal' (for example, groups of small firms), 'vertical' (suppliers and customers) and 'sectoral' (firms in the same industry) or 'lateral' (firms in different industries, but with shared interests in a technology).

J) Facilities for technology transfer: A large number of applied technology centers and other facilities to promote technology transfer have been established. These centers often extend the capabilities of existing research facilities, for example, when associated with universities, or they may be industry-driven initiatives. Centers give a visible physical presence to technology diffusion policies and house many of the activities already described, such as information provision, technology demonstration and access to new equipment, computers and software. Many countries have invested in physical infrastructure efforts to establish technology incubators and technology parks. These aim to improve links between technology developers and users through physical proximity, allowing shared access to facilities and equipment, expertise and skilled employees.

K) Regional or sectoral cluster measures: In addition to building physical facilities, governments have recognized the need to strengthen organizational capabilities and linkages within particular regions and industrial sectors, i.e., regional or sectoral clustering. Levels of communication and dialogue between technology developers and users and among users, institutional credibility and leadership as well as other aspects of 'social capital' have been shown to be extremely important in the diffusion of technology. Regional or sector cluster measures can involve strengthening industrial associations, promoting forums of stakeholders, building collaborative technology consortia, labor-management collaboration, developing leadership strategies and shared visions, along with strengthening links between users, service providers and complementary public and private assets (such as banks or training institutions). For instance, the clustering approach is gaining immense popularity especially in automotive component and light engineering industrial sectors.

L) Macro-policy measures: The overall economic and social environment has important impacts on technology diffusion. This includes factors such as business cycle stability, the cost of capital, intellectual property protection, environmental regulation, labor market policy and tax policy. A typical technology promotion measure is the offering of tax incentives or favorable depreciation schedules to enterprises that invest in new technology, new facilities or in research and development. Measures to ease regulatory burdens are also common, although in some cases tighter regulation (for example, in the environmental sphere) can promote introduction of new environmental technologies. One can observe the impact of this measure in renewable or non-conventional energy promotion by the central and regional governments through various policies and tax holidays.

6.4 Barriers to Technology Diffusion

Barriers to the diffusion of technology can be broadly classified into the following categories as explained by Cedric Philbert (2006) as below:

- Technical barriers: These barriers to diffusion of technology include the lack of competence and knowledge of planning for complex systems, adaptation or customization of the technology to suit the local conditions, possibility of using technology for various applications, availability of expert or trained manpower to implement the technology, basic infrastructure, etc.
- Economic barriers: These barriers to diffusion of technology include concerns about adoption and maintenance cost which are normally at very higher level to begin with, and impact of technology on efficiency or output. It also includes barrier like requirement of capital for updating the technology time to time.

Other barriers: These barriers include legal, cultural or behavioral barriers.

- Legal barriers like various local, national and international regulation or statutory requirements to be complied with, patents or intellectual property related issues.
- Cultural or behavioral barriers include the lack of awareness of the current status of the technology and its possibilities, anticipated side effects on environment, local culture or the livelihood. This may also include, however, the reluctance to manage a slightly more complex system or some changes in habits.

6.5 Adoption of New Technologies

It can be observed that the new technology fails more often because of failure in proper adoption and implementation of the technology rather than other reasons. The structured or disciplined project management approach may help to overcome this difficulty of failure in proper adoption and implementation of the technology. According to the Project Management Institute (PMI), a project is a work which is non-repetitive (that happens only once), has a definite beginning and end and is intended to create a unique product or knowledge. The basic management functions are planning, organizing, staffing, directing and controlling. Hence, we can define project management as 'the application of management functions, knowledge, skill, tools and techniques to project activities to meet or exceed stakeholder needs and expectations'.

The resources involved for a project like manpower, machines, money or time will vary from project to project. Also, the project may be undertaken by a single organization or consortium of organizations. The basics of project management will remain the same irrespective of the complexity of the project, i.e., whether it is a simple project of arranging a small party at home or project of building a multi-storey apartment or developing software.

The basic project management process can be summarized as follows in Figure 6.2.

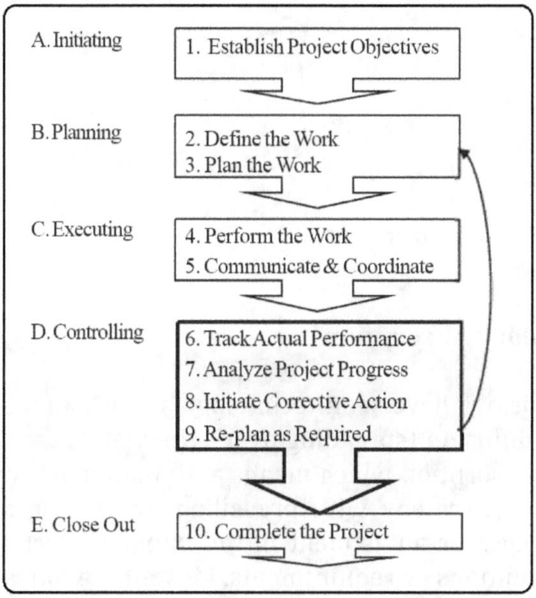

Figure 6.2 Project management process.

Each of these 10 major steps has a specific output that is defined and properly documented. This framework can help us to understand the project management in adoption and implementation of new technologies.

We can understand this concept with the help of an example of implementation of an ERP system in an organization (Figure 6.3). The decision to adopt an ERP system and subsequently implement it successfully is a very challenging task and requires tremendous efforts for any organization. Its effect has long term impact on the business and usually needs large investment in the form of money and manpower and time. It's likely to have both high potential business payback as well as threats to the business

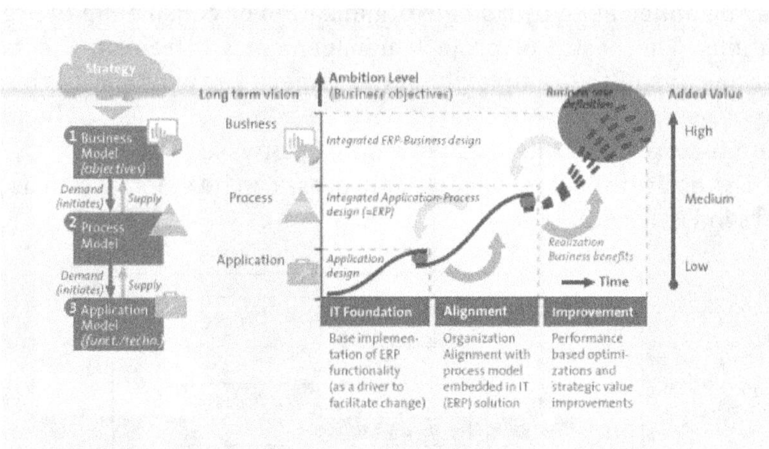

Figure 6.3 ERP implementation.

6.6 Technology Absorption

Till this topic, we have seen technology adoption (acceptance) and technology diffusion (spreading over). Now let us try to understand technology absorption which means assimilation or engrossment of technology in harmony with prevailing or existing structure, i.e., making changes or adjustment in the imported technology to suit the local conditions or requirements. Most of the times technologies are originally developed by various agencies after rigorous research in developed nations, but over the time are employed for various purposes across the globe. The technology cannot be used in an original form and requires some fine-tuning or changes for further use.

Oba & Panda (2005) termed technology absorption as 'the dynamic process involving latecomers (those who are lagging behind) in which series of new technologies are transferred from developed countries and are adopted/or adapted (assimilated) to local conditions'. As proposed by eminent economist Sanjaya Lall (1987), technology absorption is reflected in the ability to understand, deploy and improve upon technology. Oba & Panda (2005) proposed that the core of technology absorption is constituted of adoption capability and assimilation capability. The adoption capability means capability of adopting continuous-flow of new technologies in the form of new products or methods. The assimilation capability means capability of assimilating the adopted or imported technologies through learning by doing. The authors termed these two capabilities as necessary and sufficient conditions for improving productivity and product quality. The authors also opined that the role and function of human resources vary at the stage of adopting or importing new technology from the stage of absorption. At the adoption stage, the people concerned are the top managerial staff and technical experts/engineers. They focus on decision about adopting certain technology after considering definite features like:

- Technological superiority of the technology under consideration compared with the alternative and/or the existing technologies.
- Financial cost of the technology to be adopted.
- Adaptability to the local conditions (quality of raw material and labor).

At the technology assimilation stage, the concerned people are the middle level managers and workers at the workshop level. These persons are required to possess the capabilities as:

- Learning by doing skills especially by workers.
- Production and personnel management skills by managers and engineers.
- The ability to adjust to new technological changes by both workers and managers and engineers.

The technology absorption is affected by many factors such as socio-economic conditions, market forces and prevalent government regulations or policies. Especially, the market forces which include the factors affecting the price, supply, demand and availability of product along with the labor and financial factors have impact on the nature of market relations and degree of competition. The government regulations or policies include the tax structure, subsidizing, production and distribution control, pricing, etc. The technology absorption ultimately results in the improvement of product quality, cost reduction, product development as well as import substitution.

Currently numerous new technologies are developed and commercially launched in the market, but not all of them are accepted by the users immediately. It takes time for the users to understand the applicability, advantages/disadvantages of the technology along with its impact on their life. The researchers have carried out lot of studies to understand this phenomenon of new technology getting accepted, spread and assimilated in the market. In other words, we can express the same as technology adoption, diffusion and absorption. Various measures like awareness-building and technology demonstration, information search and referral services, technical assistance and consultancy, training, collaborative research and technology projects, financial support, etc., can help to increase the adoption and diffusion in small and medium scale industries. It is also observed that many barriers significantly affect the rate of technology diffusion like technology barriers, economic barriers and other barriers like legal, cultural or behavioral barriers. The technology adoption and diffusion process is a very good example of application of project management principles. Finally, to reiterate, technology absorption can be defined as 'the dynamic process involving latecomers in which series of new technologies are transferred from developed countries and are adopted/or adapted to local conditions'.

7

Technology Transfer Management

7.1 Introduction

Most of the technological invention provided mankind with lot of leisure and comforts. Some inventions are destructive, while others are dreadful such as nuclear bombs, nerve gas, etc. The importance of technology is the systematized body of knowledge and when this knowledge is put into practice, it becomes technology. Advances in science help develop new technology and new products and need for the new products drive for new specific discoveries. Technology transfer is the process by which the technology is disseminated. This will take the form of a technology transfer transaction, which may or may not be a legal binding contract but which involves the communication by the transferor of the relevant knowledge to the recipient. The successful technology transfer begins when the organization recognizes the potential of a technology for new products or process or when it recognizes a market need which is to be fulfilled. The economic benefits of an innovation or technology transfer are never fully realized until the product of the process is actually introduced in the market and cost reduction from the process is achieved. The objectives of technology transfer management are:

- To understand the inter-relationship between science and technology.
- To know how technology reaches the people through business.
- To understand impact of technology on product life cycle.
- To know how technology transfer is essential for competitive advantage.
- To understand the need of technology transfer that affects the existing industries.
- To make the job more intellectual by new technology.

Technology transfer, also called transfer of technology (TOT), is the process of transferring (disseminating) technology from the places and in groups of its origination to wider distribution among more people and places. Many companies, universities and governmental organizations now have an Office of Technology Transfer (TTO, also known as 'Tech Transfer' or 'TechXfer') dedicated to identifying research which has potential commercial interest and strategies for how to exploit it. For instance, a research result may be of scientific and commercial interest, but patents are normally only issued for practical processes. Thus, someone (not necessarily the researchers) must come up with a specific practical process. Another consideration is commercial value; for example, while there are many ways to accomplish nuclear fusion, the ones of commercial value are those that generate more energy than they require to operate. The process to commercially exploit research varies widely. It can involve licensing agreements or setting up joint ventures and partnerships to share both the risks and rewards of bringing new technologies to market.

Other corporate vehicles, e.g., spin-outs, are used where the host organization does not have the necessary will, resources or skills to develop a new technology. Often these approaches are associated with raising of venture capital (VC) as a means of funding the development process, a practice more common in the United States than in the European Union, which has a more conservative approach to VC funding. Research spin-off companies are a popular vehicle of commercialization in Canada, where the rate of licensing of Canadian university research remains far below that of the US. Technology transfer offices may work on behalf of research institutions, governments and even large multinationals. In case start-ups and spin-outs are the clients, commercial fees are sometimes waived in lieu of an equity stake in the business. As a result of the potential complexity of the technology transfer process, technology transfer organizations are often multidisciplinary, including economists, engineers, lawyers, marketers and scientists. The dynamics of the technology transfer process has attracted attention in its own right and there are several dedicated societies and journals. There has been a marked increase in technology transfer intermediaries specialized in their field since 1980, stimulated in large part by the Bayh-Dole Act and equivalent legislation in other countries, which provided additional incentives for research exploitation.

7.2 Understanding Technology Transfer

The technology transfer mainly focuses upon creating a competitive advantage for the organization. The technology transfer is focused in the international cooperation agreements between developed and developing countries. Such agreements may be related to:

- Infrastructure and agricultural development
- International cooperation in the fields of research, education, employment and transport

7.2.1 Characteristics of Technology

A) Resources:

- Money, time, people etc.
- Technology development (TD) consumes resources
- Resource may include collateral assets

B) Transferability:
Technology and knowledge transfer is not easy. Knowledge is always scarce and sticky. Between the sender and receiver during the communication, there is no perfect correspondence.

C) Opportunity:
Technology development takes place when human beings perceive an opportunity for improvement due to intrinsic or economic reasons.

D) Appropriability:
In a certain instance, technology development is for economic motives, where individuals will pursue development only to the extent that there is a reasonable assurance; the fruits of their labor will flow back to the developers. Schnepp *et al.* define technology transfer (TT) as 'a process by which expertise or knowledge related to some aspect of technology is passed from one user to another for the purpose of economic gain'. On the other hand, the Intergovernmental Panel on Climate Change (IPCC) provides a broad definition of technology transfer (TT), in terms of a set of processes, 'covering the flows of know-how, experience and equipment, for mitigating and adapting to climate change amongst different stakeholders such as governments, private sector entities, financial institutions, non-governmental organizations (NGOs) and research/education institutions'.

Some of the most often-cited considerations are listed below:

- Technology transfer includes both 'soft' and 'hard' elements of technology.
- Technology transfer should address both climate change adaptation and mitigation technologies.
- The role of government is crucial, even though the transfer of technology usually involves many stakeholders and is a complex process.
- Technology transfer should enhance the endogenous capacity of developing and helping transition countries to develop and implement change response technologies.

7.3 Process of Technology Transfer

A) Technology Management: It focuses on competitive advantages for the organization that have effective means to fulfil the needs and the requirements in the market. The methods are many. It takes a variety of forms depending on:

- Type of technology transfer
- Nature of technological assistance
- Extent of technological assistance

B) Technology Transfer: As mentioned earlier, the technology transfer is focused in international cooperation agreements between developed and developing states. Such agreements may be related to:

- Infrastructure and agriculture development
- International cooperation in the fields of research, education, employment and transport

C) Form of Technology Transfer: There are two forms of TT: internalized and externalized. Internalized technology transfer is exercised mostly by MNCs for affiliates and does not help the domestic organizations much. It refers to the investment associated with technology transfer where control resides with the technology transferor, it means the latter normally holds majority of full equity ownership. The transferor has a significant and continuing financial stake in the success of the affiliate, allows it to use its brand name and to have access to the global technology and marketing networks, exercises control over the affiliate's investment, technology and sales decisions as well as sizes the affiliate as an integral part of the global strategies. Externalized technology transfer needs lot of efforts and capital. It refers to all other forms such as:

- Joint venture and local controls
- Licensing strategic alliances
- International subcontracting
- Franchising

It may lack one or all of these features with repercussions on the technology transfer process.

D) Technology Transfer Levels:

i. Innovative level: It is the level which is characterized by innovative skills and R&D. Formal R&D is required to keep pace with technological frontiers or to generate new technologies.
ii. Operational level: It involves basic manufacturing skills. The bottom levels are the simplest one for operating a given plant which involves basic manufacturing skill, some more demanding troubleshooting, and solving problems, quality control and procurement skills along with maintenance and reliability measures.
iii. Adaptive level: Adaptive level is the self-reliance level. In this the imported technologies are adapted and improved, design skills and complexity of engineering are learned.
iv. Duplicative level: It refers to the investment capability required to expand capacity and to purchase and integrate foreign technologies at the intermediate level and duplicate the necessary skill.

E) Technology Flow Channels:

Important channels for the flow of technology are foreign investment, technology, license agreement and joint venture (Figure 7.1).

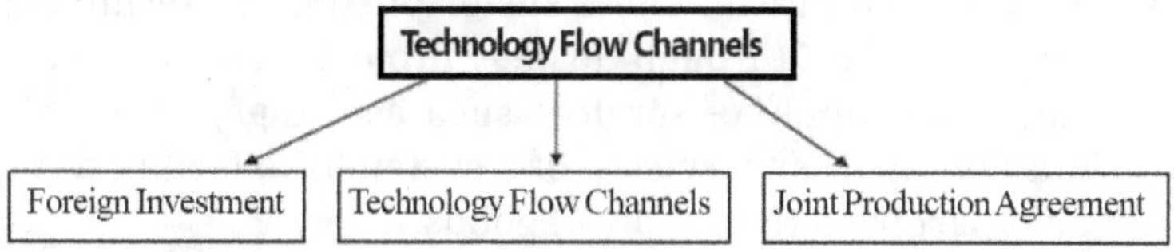

Figure 7.1 Channels of flow of technology.

F) Technology Transfer Method:

There are many methods for technology transfer and it depends on the variety of organization depending on the following:

- Type of technology
- Nature of technology assistance
- Extent of technology assistance

Following methods are normally adopted for technology transfer:

- Training of local people: Unpatented manufacturing techniques and processes can be transferred by imparting training to suitable personnel.
- Deputation of technical experts: Technology can be acquired by employing foreign technical experts for certain period. However, payment to the foreign company, subsistence allowance, transport and stay expenses, etc., are to be met.
- Licensing agreement: The licensing agreement under which the licensor or authorizing country enters into an agreement with a licensee in another country to use the technical expertise of the former. Licensee is an important means for transfer of technology. If foreign direct investment is not possible, this is the best option.
- Contract for machinery and equipment's supply: Contracts for supply of machinery and equipment which normally provide for the transfer of operational technology pertaining to such equipment is often quite adequate for manufacturing purposes not only in small projects, but also in a large number of big scale industries where the nature of technology is not particularly complex.
- Total plant engineering service contracts: After technology transfer decision is made available, complete responsibility for manufacturing and testing of product and thereafter establishing the production from the plant is handed over to the party.
- Turn key contracts: The transfer of complex technology often takes place through turnkey projects contracts which include the supply of services such as design, creation, commissioning or supervision of a system or a facility to the client apart from the supply of goods.
- Joint working arrangement: Working with the partners and

collaborators in a shared way and ensuring all the arrangements for transfer technology are made.

Many a times combination of two or more above methods are undertaken to take the best of each method.

7.4 Concerns in Technology Transfer

There are four important issues in technology transfer

- Cost
- Appropriateness
- Dependence
- Obsolescence

1) Cost: Obtaining technology from other countries involves high price. In number of cases, FDI (foreign direct investment) is associated with technology transfer. The net outflow of capital by way of dividend, interest, technical fees and royalty, etc., has been found much higher than the corresponding income flow.

2) Appropriateness: The important factor to be considered in technology transfer is physical, economic and social conditions of the developing country. In many cases, it has been found that TT has been irrelevant or inappropriate to the recipient country's socio-economic priorities and conditions. In many cases, the state-of-the-art technology is not transferred to the recipient country. It is preferable that the collaborators should offer appropriate technology by scaling down their technology at lesser cost to reduce the burden on recipient country. Most of the collaborators would think of one time benefit rather than long term benefit with the recipient country.

3) Dependence: It is imperative that the recipient country of technology transfer is heavily relying on foreign technology which leads to technological dependence on that country.

4) Obsolescence: It is normally observed that there is a tendency to transfer outdated technology to the developing countries. Therefore, the recipient country would not have the advantage of latest technology and would still technologically lag behind and normally are exploited. Many countries salvage technology that is obsolescent in the advanced country even though they possess more advanced technology.

The technology absorption in companies in the developing world has been rather slow. The main causes are:

- Lack of incentives to improve
- Resistance to change
- Lack of infrastructural facilities
- Lack of research and development (R&D) facilities
- Poor ability to absorb technology

7.5 Management of Intellectual Property Rights (IPR)

The human race, as compared to others, has a unique ability to anticipate the future and to proactively weigh and create the options for continuous survival in a dynamic environment.

There are several key technology areas, some of them are as follows:

- Smart and new materials, i.e., materials with capabilities of information storage, memory and recall.
- Bio-technology, i.e., healthcare, food, etc.
- Environmental friendly energy production and its effective utilization
- IT based infrastructure and services
- Services area of satellite industries
- Bio-engineered crops
- Gene therapy products including antigen drugs
- Bio-chip technology
- Pharmaceutical products
- Agro-chemical products

Technology creation and application coupled with enhanced cross-border access and diffusion of knowledge will dominate activities in future. It means we have to identify objective indicators of knowledge, its ownership and transfer. It is estimated that IPR now protects many products and services. The patents in addition to others form of IPR have been used as strategic business tools. The continual process of learning, creating new knowledge and further transforming into value added services with appropriate 'proprietary protection' and fair distribution of its benefits is fundamental to the progress of all knowledge based industries. Research and development generating new and high performing technologies has been given the top priority in the agenda of several national and international corporates. International 'knowledge trade' has traditionally been in the form of technology transfer from one nation to another. IPR provides formal framework of ownership for developing knowledge, for sharing of benefits between 'partners in innovation' and to create niche domains of 'value added knowledge'. These aspects form significant component of any knowledge management process in enterprises.

7.5.1 Importance of Managing IPR

Modern management of science and technology has established formal processes for innovation, value creation and sharing of knowledge within the communities for its rapid realization in the marketplace. Enhanced diffusion of knowledge has aided the rate of growth of new ideas and occupation. Formal frameworks have been set up to grant recognition to the innovator by protecting his innovation via instruments of intellectual property rights. Rules and conditions for legal use for the protected innovation have been set up and deterrents established to discourage any type of copying. IPR offers a conducive and transparent legal system for fair sharing of benefits and also to protect the interest of the consumer and of the society at large. IPR is important as industrial revolution and subsequent events enable key progress in the field of patents, industrial design registration, trademarks and copyrights. Productivity of traditional agriculture practice, food, medicines need to conserve global bio-diversity and its importance is being felt more than ever before.

The area of growing concern in the recent years has been the trade of counterfeit goods. IIPA (International Intellectual Property Alliance) and organization of US trade association are the members of these associations on copyright protection. Computer program and books account for heavy losses. The counterfeit drug results in heavy losses to the patients and the industries. The revenues from pirated CDs, etc., account for heavy loss to the industry. The business software alliance and software and information industry association (SIIS) have realized the need for a global study on software piracy and it has estimated that many software applications were pirated worldwide. The piracy-related revenue of the global software industry was estimated in billions of dollars. Such losses to businesses make it imperative to protect one's intellectual assets by using all the tools of intellectual property rights. IPR open wide possibility in legal licensing of intellectual property, creating an environment for substantiating the validity of patents, trademarks, etc.

It also encourages innovations, enhancing sharing of value added knowledge and integrating management of intellectual property in the business processes. The internet has introduced newer dimension in the field of intellectual property rights. Threats from cyber-squatters on unfair use of 'domain names', indiscriminate copying of information from diverse web sites and unauthorized distribution of copyrighted material such as recorded music videos and other software, through the cyber space by digital pirates are challenges which are required to be overcome by the inventors, owners, traders and enforcers of intellectual property. It is important that IPR management has to gear up to tackle more and more issues with the changing face of technology.

7.6 Strategic Advantage through IPR Management

The laws related to intellectual property coupled with necessary supporting legal, social, political and economic infrastructure are not as well enrooted in developing countries as in developed countries. Lack of familiarity of respect to fundamental and operating concepts of IPR in some countries contribute in enhancing the growing asymmetry/imbalance. The developing countries have to look into the equation involving enhancement of their socio-economic status, building minimum technology and management capability to derive long term benefits and harmonization process to satisfy basic requirement of TRIPS (trade related intellectual property rights). Intellectual assets preserve exclusive markets, maintain profit margin and provide access and freedom to operate. IPR portfolio has now become effective for benchmarking of intellectual assets and innovative capability of organizations. It is being used extensively in today's world of mergers, acquisition, strategic alliances, collaborations, licensing arrangements and venture capital funding. Most national IPR laws have inbuilt provisions that discourage creation of monopolistic domains and encourage legitimate completion.

National regulatory bodies exercise specific provision in the anti-trust and competition laws. Several countries have framed unilateral compulsory licensing provision in their IPR legislation that are not considered favorable in protecting the IPR of the inventors. Most IPR laws have evolved to a reasonable extent to deal with the processes used for production, however, the laws deciding on proprietorship and trade of knowledge related to them are in their rudimentary phase and need further refinement. The intellectual property rights, especially the patents for innovation in high technology area, are in the hands of developed countries. Therefore, it is evident that the primary and immediate beneficiaries of the program of the harmonization of intellectual property are likely to be the already industrialized nations. These nations have the necessary infrastructure for science and technology supported by IT based communication and information dissemination system. The modern IPR provides a range of distinct legal platforms for protection of the human creativity which spans from innovations in technical field, products and services literary/artworks, performances, computer programs, etc. Appropriate combinations of the instruments need to be exploited to achieve optimal protection of a specific intellectual asset.

Some of the strategies and the characteristic features of each of the instruments of IPR to ward off competition and increase profit through IPR management are as follows:

- Patents: It protects inventions that are novel, non-obvious with respect to prior art and useful. The invention must be disclosed in a specific format in a patent specification.
- Trade mark and service mark: Registrable trade mark such as words, signs and combinations capable of distinguishing the goods or services in connection with which they are used in the course of business helps in warding off competition.
- Intellectual property rights protection (IPR protection): Effective intellectual property rights protection promotes the smooth transfer of technology because it assures the original inventor that the technology will not be exposed.
- Trade-related aspects of intellectual property rights: The negotiations of GATT (General Agreement on Tariff and Trade) negotiations resulted in an agreement on TRIPs. Since then, the IPR has been an important part of multilateral negotiation to ward off competition and helps in increasing profits through IPR management.
- Copyrights: The Copyright Act protects the rights of the author of the original copyrightable works. The registration of work is not mandatory for the purpose of enforcement of

copyright. Hence, through registration, no new right is conferred on the owner. Through registration it only connotes an assurance that the concerned person is the genuine author. The Copyright Act 1957 is complied with the most international conventions and treaties. For instance, the Government of India has taken number of stringent actions and initiatives for the effective implementation and enforcement of copyright law. The constitution of Copyright Enforcement Advisory Council (CEAC), creation of separate cells in state police headquarters, setting up of collective administration society and organization of seminars and workshop to create greater awareness about copyright law among the enforcement personnel and general public also help in warding off competition.

The intellectual property laws in many countries are well developed. The only real problem is the implementation as the enforcement machinery is inadequate and the judicial process is slow. It's obvious that the relaxation of tariff barriers and trade liberalization have contributed greatly to the increase in flow of goods from various countries into domestic markets, thereby, intensifying competition. The abuse of IPR is now turning out to be a massive universal problem, therefore, the protection and enforcement of IPR needs a global solution. In this aspect, developing counties need to create a special task force somewhat similar to that of China for training police department all over the country in practical application of the new laws and the updated provisions in the existing laws.

8

Technology and Innovation Management

8.1 Introduction

New technology-based firms (NTBFs) are considered economically important for the renewal of traditional industrial patterns and the creation of successful innovations. Over the last few years, the rising importance of technology has led to the demand for an integral consideration of technology and innovation. Based on the integrated management theory, the 'integrated technology and innovation management' concept was developed. A distinct focus on technology management reflects, on the one side, the increasing dominance of technology and innovation in determining a company's real business viability. On the other side, it compensates for the severely underestimated impact of technological change on a company's competitiveness as described in current management literature. Technology management itself adapts to this new environmental circumstances with a shift from managing technology towards managing technology-based firms, both established and new. This means that the metaphor of technology management, being the missing link between science and technology (S&T) and general management, has to be revised. Rather than being a missing link and, thus, an activity 'outside' general management, technology management ought to be considered as an integrated part 'inside' of general management. This shift, which primarily occurs within large and multinational enterprises, converges the scope of the 'integrated technology and innovation management' in large and new technology-based firms, because in NTBF, S&T and management are automatically linked.

The term technology as a basic unit of technology management is controversially discussed in literature. This volume closely follows the definition as 'technology constitutes specific knowledge, abilities, methods and equipment, facilitating deployment of scientific and engineering knowledge'. In order to remain competitive, companies are managing technologies with four purposes:

- To enable the development of new products and services
- To allow/improve performance of specific product functions

- To serve manufacturing
- To produce products and finally to ensure companies' administrative processes and infrastructure
- The total of a company's deployed technology represents the technology potential being subjected to technology management

Within the scope of technology management, the term technology has two fundamentally different forms. Product technologies are those that deploy scientific or engineering principles to assure a specific technological impact, e.g., from optics, electronics, nuclear physics, aerodynamics, etc., deal with a specific effect and determine how an effect occurs. Process technologies, however, deploy the effects of existing product technologies to enable and optimize the occurrence of the technological impact. Research and development (R&D) process technologies are used to perform R&D activities and may include technologies such as microscopy, nanotechnology and atomic absorption technology. Typical production process technologies include galvanizing, soldering and surface mounted technology (SMT).

8.2 Managing Technologies

A technology-based firm can be conceived of as an organization which transforms input (such as components, products and services) from suppliers and partners into output (such as products, systems and services) to customers and partners. To enable this transformation, the firm has to dispose of certain assets, such as competencies and equipment. The combination and use of these assets allow competitiveness to be built up and sustained in the market place. Under this aspect, the enterprise can be viewed as an entity containing a number of specific and mostly overlapping competencies whose activation occurs by means of performing distinct processes. An enterprise finally is existentially dependent on the integration of this into its environment (Figure 8.1).

For the management of the technology enterprise, a restriction to the strategic and operational levels is unsatisfactory since factors beyond strategy play an important role. Primary among these are vision, company policy and organizational culture. This deficiency is taken into account in so-called 'integrated management' concepts, in which the strategic and operational levels are grouped under a high

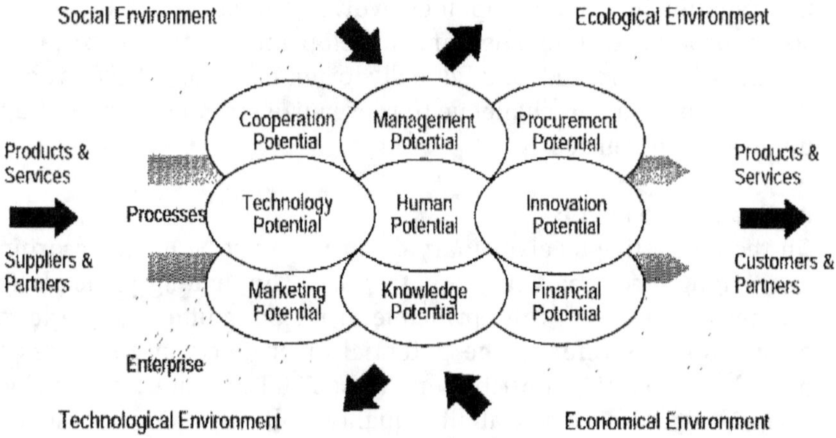

Figure 8.1 Process approach to the enterprise.

normative level of management (Figure 8.2).

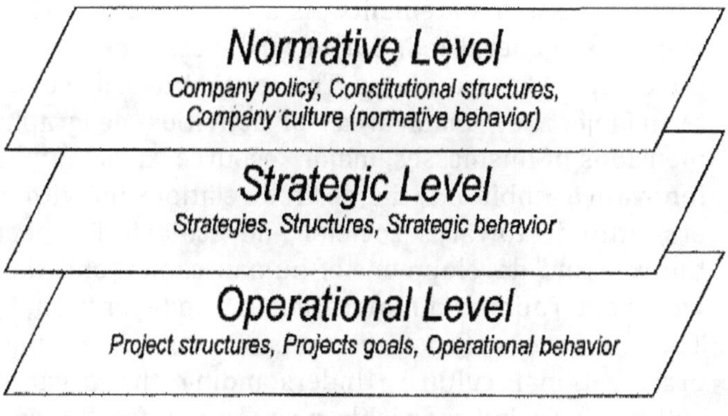

Figure 8.2 Three levels constituting general management.

These two concepts can be summarized in the framework of 'integrated technology and innovation management'. Additionally, this framework distinguishes between structural, objective-related and behavioral aspects on each management level. The formative feature of this framework is its encompassing view of technology and innovation as part of general management, which clearly shows that basically every part of the firm could be affected by technology issues. Its dynamic set-up emphasizes the fact that general management as well as technology and innovation management are always in a unique situation. Companies evolve from start-ups, by emphasizing product technologies and develop into mature enterprises, which are diversified and globally oriented. Along this process, there is a permanent challenge to reorganize structures and to build up new competencies.

A) Normative Level:

On the normative level, primary decisions must be made according to the long-term goals of the enterprise. The guiding principle for the normative level is the principle of meaningfulness. Criteria for meaningfulness refer to the potential of products and services to provide substantial contributions to societal and individual values such as organizational viability, quality of life and development of personality. Typical examples of the normative activities are as follows:

- Long-term decisions taken by company management are expressed in documents such as company vision, company policy and mission statements. As a rule, these kinds of statements are generalized, but nevertheless aim at verbalizing the company's uniqueness. The content usually covers long-term objectives, main areas of activities, geographical dimensions of businesses, major resources and competencies, innovative ambitions, the desired relationship with customers, attitude towards societal and ecological expectations, the role and development of human capital and the values which determine communication and collaboration.
- The uniqueness of each enterprise is primarily defined by its organizational culture. Understanding the organizational culture is an indispensable prerequisite for the successful leadership of an enterprise. Only cultural characteristics can ultimately explain why a new strategy has been satisfactorily implemented or not. In other words, working on a new strategy must aim at reaching a 'cultural fit', i.e., the behavioral pattern, organizational learning, group dynamics and communication capabilities.
- It is not only the making of long-term decisions which is vital for the company's future, equally essential is who takes these decisions. The far-reaching nature of technology decisions requires that original technology expertise be applied to the decision-making process.

B) Strategic Level:

On the strategic level, it is essential that company policy be transposed into comprehensive strategies. Strategies place emphasis on the selection of those resources necessary for the development and production of present and future technologies, products and services. This encompasses the question of how to acquire technologies, how to boost innovations or how to monitor relevant internal and external trends. On the strategic level, the principle of effectiveness with meaning 'doing the right thing' is prime. The following list defines and illustrates strategic technology management of structures, processes and activities as follows:

- The pattern of the development of integrated technology strategies consist of a stepwise and iterative integration of technology issues into the typical steps of strategic planning, such as setting strategic objectives, analyzing the environment, analyzing the company, elaborating strategic options, taking strategic decisions and implementing the strategy.
- Providing an adequate influx of relevant information is an enormous challenge for technology-based companies. Therefore, companies make efforts to build systems, called business technology intelligence systems, in order to keep abreast with global knowledge production. The challenge consists of building up effective scouting, screening and dissemination of capabilities in order to supply the organization with up-to-date information as an indispensable basis for taking (business) technology decisions. Staying innovative may include a careful analysis of the current and prospective innovation rate and a record of the amount of annual sales from new products. The long-range innovation rate will rise and the natural question has to be how the company is prepared to meet the prospective innovation requirements. It is in other words, the question about the appropriate content of the often cited 'pipeline'.
- Strategic technology planning implies making three fundamentally different but mutually complementary decisions, it is described as the trilogy of strategic technology decisions. The first decision originates from an analysis of current and future products. The second decision (make or buy?) is concerned with the question of whether the required technologies are to be made available through acquisition, collabora-

tion with other companies or through in-house development. The third decision (keep or sell?) deals with whether available technologies are to be applied exclusively for company purposes or can be made available to other companies.

- The task of technology marketing is to integrate acquisition and exploitation activities according to the trilogy concept. Considering the emergent technology markets, existing know-how partnerships and buy-and-sell activities gain importance to increase the firm's profitability. Traditional marketing instruments support the sales process of products, systems and services whose prices are set according to market rules and which reach the customers via real distribution channels. In contrast, technology marketing involves complex knowledge whose value is difficult to estimate, but becomes known by its reputation and its distribution takes place in the form of situational technology transfer.
- The dynamic technology portfolio is a matrix tool that provides an easily interpreted and communicated overview of current and future technology positions. Its popularity is attributed to the fact that thinking in terms of portfolios is fundamental to strategic business planning, where strategic product and business positions are to be dealt with. The strategic evaluation focuses on setting priorities as to the promotion or reduction of technology development resources. The merit of the technology portfolio lies in its high level of condensation of strategic information and at the same in its efficiency for communicating strategic decisions. In addition, a successfully finalized technology portfolio renews completion of a constructive collaboration between experts from R&D, production and marketing, which is a valuable goal on its own. In order to include technologies which are attractive despite the lack of company resources in the technology portfolio, the use of the dynamic technology portfolio is recommended instead; in addition to the traditional portfolio, as it is extended by the column 'new technologies' and at the same time by the line 'obsolete technologies'.
- Core technologies are considered a backbone of strategic technology management. Core technologies are based on core competencies which are characterized by the following features: singular, difficult to imitate, not substitutable and

valuable. In contrast to company resources which can be acquired externally, core competencies normally result from heterogenic and imperfect mobile resources. Core competence development involves organizational learning, superior cognitive capabilities and rent appropriation. As previously mentioned, core technologies are preferably original technologies developed with priority funds within the company. Though companies have to master hundreds of technologies, the number of core technologies is limited and may amount to a small proportion of all technologies, but usually give a company its unique competitive advantage.

- Management control for technology-based firms encompasses not only business but also technology strategy control. Strategic control aims to monitor the internal performance and the external environment, in order to provide a feedback on the strategy's success in the market place and to monitor the customer's and the competitor's responses. Furthermore, strategic control should check the on-going validity of the assumptions, which underlie the enterprise's strategic (technology) plan, in regard to external threats. The strategic (technology) control process itself indicates to management any deviations from the strategic plan and the necessity to take corrective actions. Potential actions during the feedback process include revising organizational strategies, re-assessing planning premises and action plans or re-thinking managerial objectives. Thus, a technology management control system should distinguish between three types of strategies: intended, emergent and realized strategies. A holistic strategic technology control approach is to upgrade Kaplan & Norton's (1992) widely accepted balanced scorecard (BSC) with a technology dimension.
- Technology roadmaps aim to effectively visualize and communicate technology forecasts and strategic technology plans. Technology road mapping is a powerful practice that supports managers taking on the challenges set by strategic technology planning. It is an integrative process that enables the organization (such as marketing, production, R&D, finance, etc.) to be aware of the problem of technology forecasting and planning. Technology road mapping enables realistic decisions to be taken more quickly and implementation to be done with more confidence and transparency.

- The technology road mapping practice emerged because the involved people such as general managers, research scientists, engineers, manufacturing mangers, marketing and finance personnel and others recognized the management problems associated with a lack of coordination and the necessity of integrating strategic technology planning.
- In order to maintain competitiveness, building optimal conditions for innovations is crucial for technology-based firms. This requires a structured evaluation of new products and new business opportunities. A product or service can be conceived of as consisting of a variety of product functions, which correspond to specific customer needs. Each product function is realized through distinct product technologies which provide a structured set of knowledge from engineering and science.

C) Operational Level:

On the operational management level, strategies are transformed into practice in the context of short-term goals, in other words personnel, technological, financial and instrumental resources are allocated to concrete R&D projects. According the principle of efficiency 'doing things right', operational management comprises the following activities as follows:

- The simultaneous engineering project management concept aims at gaining time to market through a partial overlap of individual product development phases. This implies additional risks, since essential project information may be uncertain during these overlaps. However, valuable project time may be gained, resulting in shortened R&D cycles and accelerated market entries.
- There are various cultural barriers between different units of the firm, e.g., when marketing people consider R&D people 'narrow-minded' and R&D people consider marketing people impatient, incapable of understanding technical problems and exclusively interested in short term problems. These barriers result from the natural consequence of the fact that the cultural determinants of the two groups are fundamentally different. There are three ways to cross these barriers: (1) building procedural bridges, (2) building structural bridges and (3) building human bridges.

- In order to effectively control the development of several projects, a milestone trend chart or a network diagram helps to manage the essential parameter of various projects. Product development processes are of a multidisciplinary nature incorporating several business processes, in particular R&D, manufacturing and marketing. These processes have to be coordinated and the exchange of information guaranteed.
- The Not-Invented-Here (NIH) syndrome is defined as the tendency of a project group of stable composition to believe it possesses a monopoly of knowledge of its field, which leads to rejection of new ideas from outsiders. To cope with the NIH, the changes that are likely to take place within the group as its team membership ages have to be recognized and ways to keep a project effective and performing have to be developed.
- Sharing knowledge and profiting from experience may be enhanced through databases and other documentation practices, but more important is the concept of communities of practice (CoP). A community of practice consists in general, of a team of experts working in the same field with the aim to develop their own knowledge. At the same time, the organizational knowledge base is enlarged. To ensure long-term collaboration, it is necessary that members are self-motivated. After a successful setup of a community of practice, the members should be known people, e.g., gatekeepers, to give support for general and complex problems. All activities are based on activities initiated by the core members and their particular interests.
- Reflecting quality of work in a socio-technical approach leads to an optimization of work and technology. Today's views of quality of work typically suggest a threefold integrity: physical integrity, psychological integrity and social integrity. In this context, establishing semi-autonomous working groups plays a central role, since this setting allows the group members a high degree of self-organization in terms of decision and control, autonomy of interacting with up-stream and down-stream partners and internal work organization. Another postulate is 'optimizing technology and work' meaning the development of technologies including design criteria which focus on improving work quality.

8.3 Connecting Technology and Innovation

Linking technology management and innovation management seems to be reasonable because most innovations involve the deployment of technology and technology management focuses clearly on technology and its deployment. On the other hand, innovation management extends beyond the scope of technology deployment in products and processes, e.g., it comprises social innovations which may involve new structures, processes and management approaches.

This view can be visualized using Porter's value chain: according to this widely known concept of general management, two different kinds of activities may be distinguished. On the one hand, there are those activities which are directly involved in the value generation, on the other are those contributing indirectly to value generation (Figure 8.3).

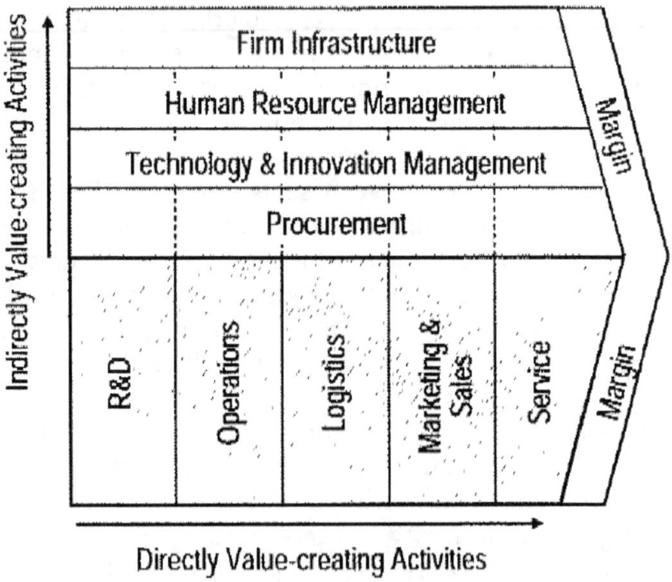

Figure 8.3 Value chain of an enterprise.

Within the value chain concept, the activities and tasks of technology management may be positioned. On the one hand, technology is deployed along the value chain via primary activities in the areas of product technology development and through logistics, production and service process technologies for products being manufactured or maintained. On the other, technology is deployed in indirect value-creating activities, i.e. infrastructural and R&D process technologies which make possible the development of product technology. The common denominator in both types of technology deployment is technology-oriented knowledge, expertise and experience, a similar picture may be presented with respect to innovation management. Innovation management covers all primary activities, including innovations which facilitate new ways of directly increasing product value. It also extends over all indirect value creating activities, including technological as well as social or business innovations. The link between technology and innovation management involves the fact that, in the majority of cases, innovations involve technology deployment, which simultaneously involves, to a large extent, innovative creations.

8.4 Technology Intelligence

It is very important to think about future trends coming from market, competitors and the technology itself. The goal of technology intelligence (TI) is the accumulation and distribution of relevant information about the firm's environment in terms of markets, competition and technology. Technology intelligence involves activities that support technological and general management by taking advantage of well-timed preparation of relevant information about technological facts and trends (opportunities and threats) in the organization's environment by means of collection, analysis and dissemination. Technology intelligence for NTBFs needs to adapt to the entrepreneurial environment, therefore, a classification in correspondence to the development of the NTB makes sense which shows four phases of the technology intelligence evolution:

- 'Primitive phase' (no specific effort)
- 'Situational phase' (awareness of the need to scan but no formal system introduced or sporadic scanning)
- 'Reactive phase' (unplanned, unstructured activities)
- 'Proactive phase' (rigorous, intensive practices)

One must also realize that NTBFs, hence also technology intelligence activities, do not evolve in a linear fashion, meaning that certain stages can be bypassed when required by a rapidly changing situation. For instance, technology intelligence can take on added importance and become more complex when the firm's environment becomes more uncertain or hostile and particularly when strong threats emerge. Other factors such as the type of competitive advantage to be obtained by a firm, its level of technological development, the quality and level of education of its leaders and its active participation in information networks can differentiate the type or level of technology intelligence done by NTBFs (Figure 8.4).

In other words, a 'primitive' or a 'situational' intelligence system can be perfectly justified if the information obtained allows the NTBF to maintain or increase its competitiveness in a specific economic environment. It is helpful to classify technology intelligence in NTBFs in four steps as a value creating process with their most dominant influential factors. However, this is not a step by step process, but a parallel assembly of diverse interacting technology intelligence activities.

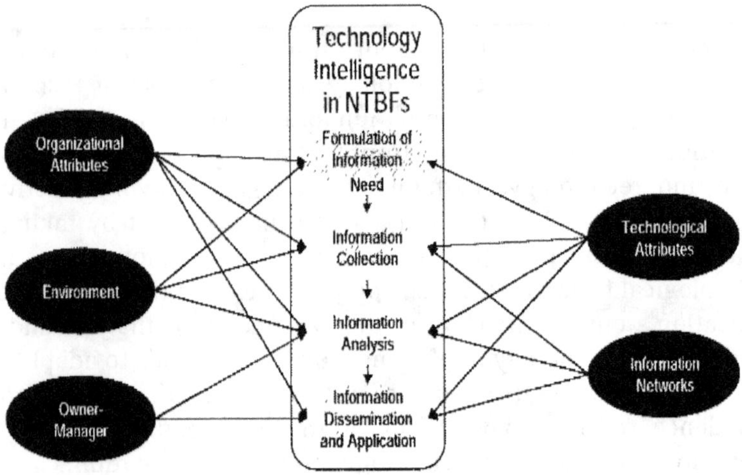

Figure 8.4 Technology intelligence.

A) Formulation of Information Need: NTBFs can have various purposes and strategic orientations when scanning and monitoring their environment. Some researchers suggest that performance is the primary motivation for scanning, while others insist on objectives related to competitiveness and strategic advantage. Various types of information needs will be formulated in an explicit or implicit formulation. An explicit formulation of an information need is normally the result of a top-down initiative. This 'input' in the technology intelligence system causes the system to react. An implicit formulation of an information need may occur as a result of emergent strategies as well as improvisational behavior.

B) Information Collection: Determining the scope of information collection is meeting the first stage, as described above. Two main questions arise up: who does it and what are the information sources? The question of who collects the information depends on its purpose. NTBFs that are doing intelligence based on implicit impulses leave it primarily to the CEO to collect the information. As soon as there are more explicit impulses, information collecting activities start being shared between different employees depending on their competencies and on availability. The assignment of the collecting task strongly depends on competencies, which can be distinguishes between internal and external technology competencies and at the same time making a distinction between formal and informal sources seems to be appropriate

C) Information Analysis: The goal of analysis is to give the information a meaning. The degree of the analysis effort depends strongly on the clarity of information and on the desired insight. Some collected information already comes to the decision-maker in an appropriate and interpreted form. Thus, no analysis is necessary. In contrast, there may be only very fragmented pieces of information which require a lot of effort to analyze. Therefore, the use of different analysis tools and the decision by whom analysis should be done depend on the context, e.g., on technology strategy (leader vs. follower) and environment complexity. Literature names different functions of the analysis stage which names three functions: filter, integration and assessment. The filter function is to reduce the quantity of information by checking the relevance of the information to the company and to assess the quality of information. This function is partly fulfilled implicitly in the collection stage. The integration function is expected to integrate information into the company context, which requires appropriate background knowledge. Then, the assessment function's role is to estimate the strategic meaning of information to the company. It is in this stage that information becomes intelligence.

D) Information Dissemination: The last dimension of technology intelligence in NTBFs refers to information dissemination and application. Information dissemination is the stage where intelligence reaches potential users, either through an information-push or information-pull channel as both channels function primarily through participation at technology intelligence meetings. Very favorable for communication seems to be transparency on 'who knows what' and there are different ways to establish the necessary transparency, for example the use of an opportunity landscape. Both, use for decision making and use for learning are key aspects of information application. Management decisions are always made when intelligence input is judged. Therefore, intelligence workers and intelligence users converge, if the intelligence tasks were shared. If technology intelligence processes are participative, organizational learning is supported by providing a useful structure.

8.5 Technology Strategy Formation

Thinking about strategies for NTBF, the saying 'one product order is worth more than thousand strategies' sets clear limits how an NTBF technology strategy has to function. There is no need for highly sophisticated strategic approaches, but every NTBF needs product orders and the technology strategy should indicate how. Therefore, the attempt of this section is to firstly define (technology) strategy and (technology) strategy formation and implementation, secondly to highlight relevant aspects for NTBF technology strategies and finally present NTBF-types of technology strategies. The technology-based enterprise finds itself in a field of tension between fundamental mid and long-term trends and discontinuities and short- and mid-term issues and uncertainties. The first includes changes in culture and social values, scientific advance and technological change as well as in industry transition, whereas the second comprises social-psychological trends and paradoxes, industry and economic fluctuations as well as changes in politics and regulatory systems. Dealing with these two types of change requires differing approaches. On the one hand, fundamental changes have to be anticipated; reacting to these changes is generally not crowned with success. On the other hand, upcoming issues are comparably easier to react to by means of strategic issues management. Therefore, managing with dual strategies requires companywide development of strategic projects for long-term concerns and competitive strategic projects for mid-term concerns. Building on the definition of strategy, a technology strategy is a plan or pattern that integrates an organization's technology goals, policies and action sequences into a cohesive whole.

This states more precisely that the purpose of technology strategies is two fold: on the one hand, technology strategies draw up a solid foundation for decision making in order to enable the selection of technologies and strategic technology fields that are suitable for the creation and maintenance of an enterprise's competitive position. On the other hand, technology strategies have to illustrate the appropriate paths leading to the mastery and deployment of the selected technologies.

8.6 Technology Management Control

Management control is divided into different subunits, one of which is technology management control. The purpose, scope and methods for strategic management control are not clearly detailed, therefore, a variety of different approaches exist in current strategic literature. The management control in NTBF plays an important role, just as much as in established firms. Comparisons between the controlling activities in NTBF and their performance showed that high performers demonstrate a superior willingness to be adaptable by modifying their business practice based on their internal evaluation and control. High performers have also established financial measures for gauging the success or failure of a new strategy, but they also considered the well-being and satisfaction of their staff as another indication of how well they were doing. For example, they would ask themselves, 'does it fulfill the financial targets and budgets we have set, and do people enjoy working here?' These firms also used fairly detailed methods of evaluating business performance in general, setting targets and comparing experience over different time periods. For example, graphs of growth, mix of products, number of information meetings held, etc., were plotted. Thus, the use of accounting information by high performing firms was acknowledged to be important and used on a regular basis to a good effect. These findings underline the importance of strategy implementation, which includes management control as an essential component. According to the definition of a high performer and the sample of firms for this study, technology plays an important role. Surprisingly, technology control was not explicitly mentioned, nevertheless, Smith (1998) concludes that for high performers:

- Total quality management systems are beneficial
- Formal quality approval gained for products, operations, personnel and business as a whole is valuable
- Regular appraisal of strengths, weaknesses, opportunities and threats must be made and reaction to those carried out
- Quality must be monitored on a regular basis, either through internal or external systems
- Regular self-critical analysis of strategy must be conducted

To introduce technology management control in NTBFs, a differentiation between a control system and control understanding is helpful. A lot of entrepreneurs cannot afford to implement complex systems. Such systems may even be dangerous because they convey a pseudo-confidence and are too detailed that they consume too many resources.

Often the implication for the gained information to management decisions is low or the management decides on information, which did not actually originate from the technology control system. In a top-down approach, a generic technology management control system is introduced, which is intended to lead at least to a better control understanding in NTBF. One suitable generic system is Simons' strategic control approach. This Technology Management Control System (TMCS) enables integration of technology aspects into management control and consists of four control levers (Figure 8.5).

	Purpose	Control Focus	Control of Technology Strategy
Technology Beliefs System	Empower and expand search activity	Vision / Mission	Perspective
Technology Boundary System	Provide limits of dissipating resources and risks	Strategic technology fields	Position
Technology Diagnostic Boundary System	Control of planned technology strategies	Plans and goals	Plan
Technology Interactive Control System	Stimulate and guide new technology decisions	Strategic uncertainties	Patterns of action

Figure 8.5 Technology management levers.

- The Diagnostic Technology Control System is the essential instrument for transforming planned technology strategy into realized technology strategy. It focuses attention on goal achievement. The diagnostic technology control system relates to technology strategy as a 'plan'. The diagnostic technology control system allows management to measure outcomes and compare results to technology goals.
- The Interactive Technology Control System is different from the diagnostic technology control system. It gives management an instrument to influence the experimentation and opportunity-seeking that may result in emergent new technology decisions. This system relates to the technology strategy as patterns of 'action'.
- The Technology Beliefs System inspires both planned technology strategy and new technology decisions. It provides guidance and inspiration for opportunity-seeking. This system relates to the technology strategy as a 'perspective'.
- The Technology Boundary System ensures that planned technology strategy as well as new technology decisions which fall within the defined strategic technology fields and are sure that activities occur at acceptable levels of technological risk. This system controls technology strategy as a 'position'.

The power of the technology control levers is not only located in how each is used alone, but rather in how they complement each other when used together. These interplays create a dynamic tension between the opportunistic innovation and predictable goal achievement that is necessary for an enterprise's technology performance. The interplay of different technology control levers enables management to control the enterprise's technology portfolio as well as the technology strategy and its performance.

9

Issues in Technology Management

9.1 Introduction

The social and cultural environment is governed by the conditions including customs, morals, values and demographic characteristics of the society in which the technology transfer has taken place dealing with different expectations of people. The social and cultural environment is the crux of cross cultural management and to be effective in this area, the organization will need to develop a broader understanding of the social environment when technology transfer is required. The business ethics demands a close analysis of these prerequisites as also the social and business structures that are conducive to moral principles. Right from the beginning, the advances and innovations in technology and sciences are interconnected with the social issues of that age. Every field of endeavor must come to terms with the social implication of these advances and innovations. No research topic of social significance can afford to ignore the technology or science. These effects are consequential more acutely than ever. The social issues are more environmentally based but that is indicative of contemporary world. Issues such as nuclear power plant, hazardous wastes and examining human gene therapy, the search for extraterrestrial intelligence, etc., are social issues. It uses three part approaches:

- Brief background on the technology and science
- Historical contact showing social, ethical, economical and other issues evolved
- Various points of view on each issue

The business and society have a symbiotic relationship, with business acting the role of the junior partner. It is strongly affected by the surrounding social, political, ethical, moral and ecological environment. Society exerts influence on business in several ways:

- The society legal system
- Political and government regulations

- Public opinion and community attitudes
- Influence on nature of business, growth and development
- The concept of morality and ethics held in that particular society determines its development

9.2 Social Issues

The advancements in science and technology have always been associated with the social issues. Whether technology has contributed to the question about a specific problem, or has been its cause, new scientific discoveries and technologies never get away from inspection. Every field of venture must come to provisions with the social implications of these advancements. Similarly, anyone researching on a topic of social significance cannot overlook the science or technology, if present, which plays a part in it. In our modern culture, these effects are felt more intensely than ever.

Following are the social issues related to technology in specific areas:

- **Survival**: The advancement in technology is mostly used for destruction purposes as technology does not have a persistent ethical code to which people almost commonly subscribe or effective means of detecting and policing breaches. In the light of these issues, the survival becomes an important issue.

- **Use of Internet**: Internet is a universally accessible technology. It is being used along with computer. As we all know that computer and internet has become one of the basic requirements for us. Internet can be used for both constructive and destructive purposes. This becomes a social issue for the business & the government.

- **Employment (Work) Related Concerns**: All of us should think about the answer to the question 'how does a less specialized society offer everyone of us with job?' The problem solvers will always be working, however, only the type of the problems they solve will alter. For example, robotics guarantee that there will be few unskilled or repetitive jobs in the future civilization. However, they will not take over complete work until and unless we design them to do so. All these concerns provide a professionalization of work and not its elimination.

- **Artificial Intelligence**: This is an interesting field regarding the use of technology which is being used to replace various human activities, enhancing human skills, etc. The purpose behind this is its use for a noble cause.

- **Environment Related Concerns**: Environment is very important not only for human society, but also for the whole living being on the globe. The technology has to be environment friendly. Technology has helped humankind to discover the nuclear power which is both a boon and bane to us. The products and wastes while generating nuclear power are harmful for the environment.

Due to technological developments, many issues are arising, issues related to social and human welfare, urban and regional challenges, education and employment. The issues related to socio-demographic and technology-transfer are:

- Problems associated to an aging population and the implementation of new technology
- Socially oriented technologies
- Surveys indicating varying demographic arrangement and the position of technical education
- Effect of new technologies on human behavior

Social issues in science and technology are variable and provide a good overview and valuable background information about a variety of topics.

9.3 Ethical Issues

There are various ethical issues in relation to technology management that need to be addressed. The organizations are testing the system or performing an immoral action or will genetic engineering improve the quality of peoples' lives or start destroying it? How do we recognize when an ethical issue exists?

There are, indeed, many grey ethical areas. The important ethical issues are as follows:

A) Plagiarism: Plagiarism is where the work of others is copied, but the author presents it as his or her own work. This is a highly unethical practice, but happens quite frequently, and with all the information that is now available on the internet, it is much easier to do and is happening more often. As Bowyer states: Plagiarism is the taking of the ideas, writings, drawings, words or other similar intellectual property created by others and presenting it as your own. It is generally not a legal issue, like copyright infringement, but it is an ethical one. For example, you can reuse writings in the public domain without worrying about the legal problem of infringing a copyright, but presenting them as your own without proper credit to their true origin is an act of plagiarism and plagiarism is unethical. Bowyer also refers to 'self-plagiarism', whereby the author reuses his/her own words from a previous publication in a newer publication without referencing the older publication.

B) Piracy: Piracy, the illegal copying of technology, is a very serious problem. Although some might argue that some pirating at least should be permitted as it can help in leading to be more usefulness for business. However, for corporations, in particular, this is a very serious issue, and can significantly damage profit margins.

C) Ergonomics/Health Issues: Ergonomics, also known as human factors, is the scientific discipline that seeks to understand and improve human interactions with products, equipment, environments and systems. There are many ergonomic/health issues related to technology. Responsible/ethically-minded employers will, hopefully, give due consideration to this, as indeed should all employers. As Kallman and Grillo say: Ergonomics is concerned with the physical work environment. The question is how far should an organization go to be 'ergonomically sound'? For example, what is required to provide data entry clerks with a healthful work area? How can a firm create an environment that results in minimal eye strain, guards against back problems, prevents repetitive-motion syndrome and protects against exposure to possibly harmful CRT (cathode-ray tube) emissions? Without such ethical/moral awareness and taking the necessary action, many workers will suffer health problems directly from new technology.

D) Job Displacement/Work Pressures Imposed on Personnel: Due to technology upgradation, employees' jobs are becoming redundant quite often or they have to play quite different roles and some are suffering increasing levels of stress from work pressures. Some are, obviously, reaping the benefits of having more rewarding jobs and there is certainly more emphasis on knowledge and information. There are many ethical issues that need to be addressed.

E) Digital Divide: The digital divide poses a serious problem today. A new breed of 'have's' and 'have nots' are being created, between those that have access and can use a technology computer and those that do not have such access. There are clearly serious ethical implications here. Those that do not have such access may well be discriminated against, feel 'socially excluded' and miss out on many life opportunities.

F) Gender Bias: There are also ethical issues in regards to gender and technology, given the fact that females are often discriminated against in various ways in this era of new technology.

G) Issues related to Nanotechnology: Nanotechnology presents a new set of ethical dilemmas. As Colvin says: For the past decade, nanotechnologists have basked in the glow of positive public opinion. We've wowed the public with our ability to manipulate matter at the atomic level and with grand visions of how we might use this ability. All this 'good news' has created a growing perception among business and government leaders that nanotechnology is a powerful platform for twenty-first technologies (Colvin, 2002).

9.4 Moral Issues

Morals or the idea of morals presupposes the idea of social rules and are concerned with the judgment of the goodness or badness of human action and character. The moral rules are NOT enforced or legislated by a specific body of people designated or appointed. For that purpose, morality is informal. The moral issues confronted in technology management are as follows:

A) Environmental Protection

Environmental protection comprises of the following:

- The environmental protection is concerned with human-environment interaction.

- Concern for future generation
- Pollution activities
- Human interests and eco-system

B) Consumer Safety

A product is more than just the physical object. It is the pride of the organization. It interacts with customers and has following issues:

- Product safety
- Product standards
- Manufacturers' obligation to inform customers

C) Question of Loyalty

In case an employee perceives an organizational practice as illegal/unethical, he seeks to correct it by bringing it to the knowledge of the competent authorities in the organization and in case of no result, he notifies the authorities outside the organization. This practice is known as 'whistle blowing'. It has following three categories:

- Internal whistle blowing
- Governmental whistle blowing
- Corporate whistle blowing

Whistle blowing may cover a wide range of activities dissimilar from a moral point of view.

9.5 Human Aspects in Managing Technology

We are well aware of the concepts of technology management and various aspects related to technology assessment and strategy, technology transfer, forecasting, etc. However, all of these require human intervention which plays a very vital role in technology management. There is hardly any facet of life which is untouched by technology. Technology may be very sophisticated and state of the art, but its successful deployment needs human intervention or involvement. Apart from individual, human-technology interaction takes place meticulously at organizational level than in personal life. Hence, while understanding the human aspects of technology management, we will explore this topic here mainly from the organization point of view. We have come a long way from the agrarian era where farming was the mainstay of life to the information technology era of the 21st century. The context of life and employment changed dramatically after the industrial revolution in the 18th century. The industrial revolution transformed the agricultural and manufacturing practices forever. It also had endearing effect on other facets of humanity such as transportation and mining. One of the most important drivers of industrial revolution was to improve production efficiency, i.e., to manage more output with less input, especially manpower. During the industrial revolution, the introduction of various machines triggered the rise in the production efficiency and had an intense effect on the employment patterns.

The industrial revolution is succeeded by information technology revolution which has spread faster than its predecessor. Almost everything in our life today is affected by information technology. The factors like intense competition and rapidly changing customer as well as stakeholder's demands put pressure on various organizations to embrace new technologies. In turn, the human resources are under the pressure to update their knowledge and skill to sustain the changes effected by these new technologies. The technology is significantly affecting the business practices and the human resource management which is becoming complex. The role of human aspect in technology management is a multi-disciplinary task, i.e., it has to be studied from different perspectives like psychology, sociology, economics, engineering, statistical sciences, etc. Understanding various aspects of technology and its successful implementation for the constructive purpose depends on manpower.

The role of human aspect in managing technology can be understood in two ways:

1) As one of the enabler in innovation or research and development of technology continuously improving the industrial or business practices: As mentioned by Zuboff (1988), technology occupies a critical position in the development of organizations and nations as lack of technology makes organizational competitiveness and national progress difficult. According to Yunus Dauda (2008), technology development is based on the outcome of basic scientific findings that were exploited to bring out specific needs. This includes fabrication of equipment for the improvement of methods, steps and processes necessary to bring out new or modified products or services into market for human needs. Specific machines and processes are provided through science and technology to improve performance and it is possible through continuous innovation. Technological innovation requires well-trained, motivated, experienced and dedicated manpower. It ultimately means effective human resource management is crucial for organizational effectiveness and competitiveness.

2) As an important element in implementation of innovations or research and development findings (usually in the form of new/updated technology): The innovations or research development will usually enable new technologies which may be state of the art, but human aspect plays a very decisive role in its success or failure. 'Simply put, nine system failures out of ten, it's the human factor, and even that 10th example probably has a very strong human component hiding somewhere', mentions Allen Bernard (2003). Given that a system is designed, built, commissioned, operated, maintained, modified and decommissioned by humans, a root-cause analysis can always find a human factor somewhere as opined by Ned Hickling.

A framework of technology based on task interdependence proposed by James Thompson in 1967 classifies types of technology into three types as: mediating, long-linked and intensive technologies. It is, focusing on technology as a dimension to understand the actions of complex organization.

A) Mediating Technology: Mediating technology is based on pooled task interdependence. Organizations using mediating technology would have organizational tasks being carried out independently and each task's performance contributing separately to the performance of the whole organization. Thus, here the task interdependence is quite low as performance of a task does not require direct reliance on other tasks.

B) Long-Linked Technology: Long-linked technology involves sequential interdependence between tasks whereby the output of the preceding task becomes the input for the succeeding task. Thus, the performance of the succeeding task depends on the performance of the preceding task and cannot be performed before the preceding task is completed. Therefore, there is need to manage coordination between the sequential tasks for effective development of the final output in long-linked technology.

C) Intensive Technology: Intensive technology involves reciprocal interdependence between the various tasks such that performance of one task is dependent on the performance of the other tasks and vice versa. Organizations using intensive technology would need to select, combine and apply tasks based on the feedback received from the output produced. Thus, the overall output of intensive technology would depend on the nature of coordination achieved between the tasks. Undoubtedly in the current era, information and communication technology (ICT) has affected the organization structures like never before.

Why use of ICT is increasing in organizations is the topic of separate debate, but the reasons are obvious and linked to overall productivity and efficiency of the organizations. Bresnahan *et al.* mention that the complementarities between machine and human capabilities lead to a wide variety of organizational structures which includes information-enabled decentralized worker. It involves support of analytical applications enabling the worker to know the things necessary to successfully carry out the job. The computer based analytical application permits the superiors of the worker to monitor his/her performance quickly and efficiently. Other concepts which we should briefly consider while understanding the relationship between technology and organization structure are 'virtual teams' and 'virtual organization'.

9.6 Environmental Impact Analysis

An environmental impact analysis (EIA) is the analysis of possible positive or negative impact that a proposed activity may have on the environment. It also consists of the social and the economy aspects. Environmental impact analysis is defined as 'the process of identifying, predicting, evaluating and mitigating the biophysical, social and other relevant effects of development proposals prior to major decisions being taken and commitments made'. Environment impact analysis is unique as it does not require adherence to a predetermined environmental outcome, rather it requires decision makers to account for environmental values in their decision and to justify those decisions in the light of detailed environmental studies and public comments on the potential environmental impact of the proposal. The technological environment can be thought to be a subset of systems of social philosophies, norms, dogma-system of belief, ideologies, etc. The environment consists of the task, industry and macro environment. The macro environment consists of social, economic, political, regulatory and technological elements. These are the environment levels.

The environmental impact analysis essentially is to determine, what the project is designed to do and how the project will affect the environment both in the long and short term and what the alternatives to the projects are. Normally, such studies require the work of cooperating team of engineers, planners, and scientists such as geologist, chemists, hydrologists and biologists, depending upon the type of projects. The purpose of environmental impact analysis is to ensure that the decision maker considers the ensuing environmental impact when deciding whether to proceed with a project or not.

EIA involves a technical evaluation that would lead to objective decision making. EIA as it is practiced today is being used as decision aiding tool rather than decision making tool, there is growing dissent on the use of EIA as its influence on development decision is limited and there is a view that it is falling short of its full potential. Broadening the scope of EIA can be beneficial. Instead of concentrating on the direct effects of a proposed project on its local environment, the benefits of EIA can be used as a landscape approach which is focused on much broader relationships between the entire populations of a variety of species. As a result, an alternative would cause least amount of negative effects to the population of the species as a whole, which is the greatest benefit of EIA. EIA can help in making R&D of nation successful and increase productivity of in-house R&D. It helps in deployment of quality principles which, in turn, helps in strategic decision making, administrative and research processes which include the exception of client's evaluation of R&D process and reviews. The cost functional integration and management includes inter-functional integration of R&D with manufacturing, marketing and engineering as well as reduction in time to delivery, along with control of projects. It also helps participation of scientists in the design and execution of research programs and training of scientists. EIA also helps the companies to establish R&D operations responsive to national market and to cooperate with various laboratories. It also serves as a window for the development in the respective region.

References

1. Barley, S. R. (1986) Technology as an occasion for structuring: Evidence from observations of CT scanners and the social order of radiology departments. *Administrative Science Quarterly*, **31**(1), 78-108.
2. Bhalla, S. K. (1987) *The Effective Management of Technology*, Batelle Press.
3. Bresnahan T. F., Brynjolfsson E., and Hitt L. M. (1999) Information Technology and Recent Changes in Work Organization Increase the Demand for Skilled Labor. Available from: http://www.stanford.edu/~tbres/research/Brookings%20volume.pdf.
4. Burgelman R. A., Christensen, C. M., and Wheelwright, S. C. (2001) *Strategic Management of Technology and Innovation*, McGraw-Hill.
5. Agalgatti, B. H., and Krishna, S. (2014) *Business Ethics*, Nirali Prakashan.
6. Chiesa, V. (2001) *R&D Strategy and Organization*, Imperial College Press.
7. Davidow, W. H., and Malone, M. S. (1992) *The Virtual Corporation: Structuring and Revitalizing the Corporation for the 21st Century*, Harper Business.
8. Drucker, P. F. (1989) *The Practice of Management*, Heinemann Professional.
9. Dublin, M. (1989) *Futurehype: The Tyranny of Prophecy*, Plume.
10. EIA Manual of Ministry of Environment and Forests (MoEF), Government of India. Available from: http://moef.nic.in/divisions/iass/eia/Cover.htm.
11. Ford, D., and Saren M. (1996) *Technology Strategy for Business*, Thomson Business Press.
12. Gaynor, G. H. (1996) *Handbook of Technology Management*, McGraw-Hill Professional.
13. Hall, B. H., and Khan, B. (2002) Adoption of New Technology, New Economy Handbook. Available from: http://elsa. Berkeley.edu/~bhhall/papers/HallKhan03%20diffusion.pdf.
14. Prahalad, C. K., and Hamel, G. (1990) The core competence of corporation. *Harvard Business Review*, May-June issue.
15. Handy, C. (1995) Trust and the virtual organization. *Harvard Business Review*, May-June issue.

16. Hanke, J., and Reitsch, A. (1992) *Business Forecasting*, 4th edition, Simon & Schuster.
17. Hax A. C., and Majluf A. S. (1996) *The Strategy Concept and Process: A Pragmatic Approach*, Prentice Hall.
18. http://moef.nic.in/modules/others/?f=eia-manuals
19. http://techtransfer.energy.gov
20. http://www.bing.com/images/search?q="technology+life+cycle"
21. http://www.financialexpress.com/news/india-to-formulate-new-science-and-technology-policy/563207/
22. http://www.investopedia.com/terms/s/standardization.asp#axzz1Zsj68jd0
23. http://www.tdb.gov.in/WebContent.aspx?id=10&type=homemore
24. Pal, P. (2008) *Intellectual Property Rights in India*, Regal Publication.
25. Internal Factors Affecting Organisations. Available from: http://www.strategy-formulation.24xls.com/en125.
26. Ashton. T. S., and Hudson, P. (1998) *The Industrial Revolution*, Oxford University Press.
27. Makridakis, S., Wheelwright, S. C., and Hyndman, R. J. (1998) *Forecasting: Methods and Applications*, 3rd edition, John Wiley.
28. Murthy, C. S. V. (2010) *Management Information System*, Himalaya Publishing House.
29. Khalil, T. M. (2012) *Management of Technology: The Key to Competitiveness and Wealth Creation*, Tata McGraw Hill.
30. Babcock, D. L. (1991) *Managing Engineering and Technology: An Introduction to Management for Engineers*, Prentice Hall.
31. Dauda, Y., (2008) *Managing Technology Innovation: The Human Resource Management Perspective*, Peter Lanh GmbH.
32. Martino, J. P. (1983) *Technological Forecasting for Decision Making*, 2nd edition, McGraw Hill.
33. Porter, M. E. (1985) *The Competitive Advantage: Creating and Sustaining Superior Performance*, Free Press.
34. Millett, S., and Honton, E. (1991) *A Manager's Guide to Technology Forecasting and Strategy Analysis Methods*, Battelle Press.
35. Information available at the website of Ministry of Science & Technology (Government of India).
36. Mulder, P., de Groot, H. L. F., and Hofkes, M. W. (2003) Explaining slow diffusion of energy-saving technologies; a

vintage model with returns to diversity and learning-by using. *Resource and Energy Economics*, **25**(1), 105-126.

37. Mulins, L. (2007) *Management and Organisational Behaviour*, 7th Edition, Pearson Education.
38. Hickling, N. (2010) Human Factors and Failure – A Viewpoint on Risk Perception. Available from: http://analysis.nuclearenergyinsider.com/qa/human-factors-and-failure-%E2%80%93-viewpoint-risk-perception.
39. Harrison, N., and Samson, D. (2001) *Technology Management: Text and International Cases*, McGraw-Hill/Irwin.
40. How Farm People Accept New Idea, Special Report No. 15, North Central Committee Regional Publication No. 1 of the Agricultural Extension Services, Iowa State University, 1955. Available from http://www.soc.iastate.edu/extension/presentations/ publications/comm/SP15.pdf.
41. Orlikowski, W. J. (1992) The Duality of Technology: Rethinking Concepts of Technology in Organizations. Available from: http://links.jstor.org/sici?sici=10477039%28199208%293%3A3%3C398%3ATDOTRT%3E2.0.CO%3B2-U.
42. Philibert, C. (2006) Barriers to Technology Diffusion: The Case of Solar Thermal Technologies, IEA Working Paper Series, IEA/OECD, Paris.
43. Chase, R. B., Aquilano, N. J., and Jacobs, F. R. (1998) *Production and Operations Management: Manufacturing and Services*, McGraw Hill.
44. Rosenberg, N. (1972) Factors affecting the diffusion of technology. *Explorations in Economic-History*, **10**(1), 3-33. Reprinted in Rosenberg, N. (1976) *Perspectives On Technology*, Cambridge University Press.
45. Schein, E. H. (1988) *Organizational Psychology*, 3rd edition, Prentice Hall.
46. Science Digest magazine.
47. Self-analysis based on work experience.
48. Shapira, P., and Rosenfeld, S. (1996) An Overview of Technology Diffusion Policies and Programs to Enhance the Technology Absorptive Capabilities of Small and Medium Enterprises, a background paper for the Organization for Economic Cooperation and Development (OECD), Directorate for Science, Technology and Industry, Paris. Available from: http://www.prism.gatech.edu/~jy5/pubs/oecdtech.htm.

49. http://techtransfer.energy.gov
50. Technology Life Cycle, Tera Sigma Consulting.
51. Murthy, C. S. V. (2012) *Technology Management*, Himalaya Publishing House.
52. The Virtual Organisation, http://www.economist.com/node/14301746.
53. Thompson, J. D. (1967) *Technology-Task Interdependence Framework: Organizations in Action*, McGraw Hill.
54. Tidd J., and Bessant, J. (2001) *Managing Innovation: Integrating Technological, Market and Organizational Change*, John Willey.
55. Twiss, B. C. (1992) *Forecasting for Technologists and Engineers: A Practical Guide for Better Decisions*, Institution of Electrical Engineers.
56. Narayanan, V. K. (2009) *Managing Technology and Innovation for Competitive Advantage*, Pearson Education.
57. Virtual Organisation definition as available on http://sensacom.com/web_glossary.html and http://whatis.techtarget.com/definition/0,,sid9_gci213301,00.html.
58. Kinicki A., and Kreitner, R. (2008) *Organizational Behaviour: Concepts, Skills and Practices*, 2nd edition, McGraw-Hill.
59. Woodward, J. (1965) *Industrial Organization: Theory and Practice*. Available from: http://www.unc.edu/~nielsen/soci410/nm6/m4002.gif.
60. Zuboff, S. (1988) *In the Age of the Smart Machine*, Basic Books.

List of Important Words, Terms and Phrases

- Accountability: The mechanisms for assessing responsibility for decisions made and actions taken.
- Agency management: A term that refers to those people who are responsible for the business operations of an agency.
- Authorization: The power granted by management to specified individuals allowing them to approve transactions, procedures or total system.
- Balanced scorecard: Model for analyzing firm performance which supplements traditional financial measures with measurements from additional business perspectives, such as customers, internal business processes, and learning and growth.
- Batch processing: A method of collecting and processing data in which transactions are accumulated and stored until a specified time when it is convenient or necessary to process them as a group.
- Benchmark: A measurement or standard that serves as a point of reference from which process performance is measured.
- Benchmarking: A structured approach for identifying the best practices from industry and government, and comparing and adapting them to the organization's operations. Such an approach is aimed at identifying more efficient and effective processes for achieving intended results and suggesting ambitious goals for program output, product/service quality and process improvement.
- Business continuity planning: Planning that focuses on how the company can restore business operations after a disaster strikes.
- Business intelligence: Applications and technologies that focus on gathering, storing, analyzing and providing access to data from many different sources to help users make better business decisions.
- Business model: An abstraction of what an enterprise is and how the enterprise delivers a product or service, showing how the enterprise creates wealth.
- Business process management: Methodology for revising the organization's business processes to use these as fundamental building blocks of corporate information systems.

- Business process reengineering: The radical redesign of business processes, combining steps to cut waste and eliminating repetitive, paper-intensive tasks in order to improve cost, quality and service as well as to maximize the benefits of information technology. In government, it represents a systematic disciplined improvement approach that critically examines, rethinks and redesigns mission-delivery processes and sub-processes within a process management approach. In the political environment, the approach achieves radical mission performance gains in meeting customer and stakeholder needs and expectations.
- Business processes: The unique ways in which organizations coordinate and organize work activities, information and knowledge to produce a product or service.
- Change management process: A set of tasks or procedures established to ensure that project performance is measured to the baseline and changes are reviewed, approved or rejected and the baseline updated.
- Configuration management: Configuration management is a formal discipline that provides project team members and customers with the methods and tools that are used to identify the developed product, establish baselines, control changes to these baselines, record and track status and audit the product.
- Contingency planning: The development of a management plan that identifies alternative strategies to be used to ensure project success if specified risk events occur.
- Control system: A mechanism that reacts to the current project status in order to ensure accomplishment of project objectives.
- Core processes: Processes that have clear dependencies and that require the same order on most projects.
- Cost of quality: Determining the costs incurred to ensure quality.
- Critical path: Generally, but not always, the sequence of schedule activities that determines the duration of the project. Generally, it is the longest path through the project. However, a critical path can end, as an example, on a schedule milestone that is in the middle of the project schedule and that has a finish-no-later-than imposed date schedule constraint.

- Critical success factors: The limited number of areas of performance that are essential for a project to achieve its goals and objectives. These are the key areas of activity in which favorable results are absolutely necessary to reach goals.
- Delphi technique: An information gathering technique used as a way to reach a consensus of experts on a subject. Experts on the subject participate in this technique anonymously. A facilitator uses a questionnaire to solicit ideas about the important project points related to the subject. The responses are summarized and are then re-circulated to the experts for further comment. Consensus may be reached in a few rounds of this process.
- E-government: Use of the internet and related technologies to digitally enable government and public sector agencies' relationships with citizens, businesses and other arms of government.
- Enterprise technology program: A group of related IT projects aggregated for management purposes that support a defined enterprise.
- Information system: Interrelated components working together to collect, process, store and disseminate information to support decision making, coordination, control, analysis and visualization in an organization.
- Knowledge network: Online directory for locating corporate experts in well-defined knowledge domains.
- Local area network: A telecommunications network which requires its own dedicated channels and encompasses a limited distance, usually one building or several buildings in close proximity.
- Net present value: The amount of money an investment is worth, taking into account its cost, earnings and the time value of money.
- On-line analytical processing: Capability for manipulating and analyzing large volumes of data from multiple perspectives.
- Organizational impact analysis: Study of the way a proposed system will affect organizational structure, attitudes, decision making and operations.
- Organizational learning: Creation of new standard operating procedures and business processes that reflect organizations' experience.

- Organizational memory: The stored learning from an organization's history that can be used for decision making and other purposes.
- Outsourcing: The practice of contracting computer center operations, telecommunications networks or applications development to external vendors.
- Post-implementation audit: Formal review process conducted after a system has been placed in production to determine how well the system has met its original objectives.
- Predictive analysis: Use of data mining techniques, historical data and assumptions about future conditions to predict outcomes of events.
- Present value: The value, in current dollars, of a payment or stream of payments to be received in the future.
- Radio frequency identification: Technology using tiny tags with embedded microchips containing data about an item and its location to transmit short-distance radio signals to special RFID readers that then pass the data on to a computer for processing.
- Supply chain management: Integration of supplier, distributor and customer logistics requirements into one cohesive process.
- Systems analysis: The analysis of a problem that the organization will try to solve with an information system.
- Systems design: Details how a system will meet the information requirements as determined by the systems analysis.
- Systems development: The activities that go into producing an information systems solution to an organizational problem or opportunity.
- Systems lifecycle: A traditional methodology for developing an information system that partitions the systems development process into formal stages that must be completed sequentially with a very formal division of labor between end users and information systems specialists.
- Total quality management: A concept that makes quality control a responsibility to be shared by all people in an organization.
- Value chain model: Model that highlights the primary or support activities that add a margin of value to a firm's products or services where information systems can best be applied to achieve a competitive advantage.

Index

A

B

C

D

E

F

G

I

L

M

P

Q

R

S

T

V

www.ingramcontent.com/pod-product-compliance
Lightning Source LLC
LaVergne TN
LVHW022044200926
848481LV00003B/785